Computer Simulated Experiments for Digital Electronics Using Electronics Workbench®

Richard H. Berube
Community College of Rhode Island

Prentice Hall

Upper Saddle River, New Jersey   Columbus, Ohio

Cover art: Interactive Image Technologies, Ltd.
Editor: Linda Ludewig
Production Editor: Rex Davidson
Composition: Brenda Dobson
Cover Designer: Karrie M. Converse
Production Manager: Patricia A. Tonneman
Marketing Manager: Ben Leonard

This book was printed and bound by Courier/Kendallville. The cover was printed by Phoenix Color Corp.

Printed in the United States of America

10 9 8 7 6 5 4 3 2

ISBN: 0-13-749475-0

Prentice-Hall International (UK) Limited, *London*
Prentice-Hall of Australia Pty. Limited, *Sydney*
Prentice-Hall of Canada, Inc., *Toronto*
Prentice-Hall Hispanoamericana, S. A., *Mexico*
Prentice-Hall of India Private Limited, *New Delhi*
Prentice-Hall of Japan, Inc., *Tokyo*
Editora Prentice-Hall do Brasil, Ltda., *Rio de Janeiro*

Preface

Computer Simulated Experiments for Digital Electronics Using Electronics Workbench® is a unique and innovative laboratory manual that uses Electronics Workbench® to simulate digital laboratory experiments on a computer. Computer simulated experiments do not require extensive laboratory facilities, and a computer provides a safe and cost-effective laboratory environment. The digital circuits can be modified easily with on-screen editing, and analysis results provide faster and better feedback than a series of experiments using hardwired digital circuits. The materials list and circuit diagrams included with each experiment make this manual also usable in a hardwired laboratory environment, if hardwired experience is desired.

The experiments in this manual are designed to help reinforce the classroom theory learned in a digital electronics course and can be used with any digital textbook. By answering questions about the results of each experiment, students will develop a clearer understanding of the theory. Also, the interactive nature of these experiments encourages student participation, which leads to more effective learning and a longer retention of the digital concepts.

The experiments in Part I of this manual involve the study of logic gates and combinational logic circuits. The experiments in Part II involve the study of arithmetic logic circuits such as binary adders, BCD adders, parity generators and checkers, and magnitude comparators. The experiments in Part III involve medium scale integrated (MSI) circuits such as decoders, encoders, multiplexers, and demultiplexers. The experiments in Part IV involve the study of sequential logic circuits such as latches, flip-flops, monostable and astable multivibrators, registers, and counters. The experiments in Part V involve the study of circuits that interface the digital world with the analog world for the acquisition of data.

An experiment with troubleshooting problems is included at the end of each section to help students develop troubleshooting skills. In each troubleshooting problem, the parts bin has been removed to force the student to find the fault by making a series of circuit measurements rather than by replacing components. A solutions manual showing measured data, answers to the questions, and answers to the troubleshooting problems is available.

I wish to thank the following reviewers for their valuable suggestions: Michael A. Agina, Texas Southern University; V. S. Anandu, Southwest Texas State University; L'houcine Zerrouki, ITT Technical Institute; Julio Garcia, San Jose State University; Byron Paul, Bismarck State College. I also wish to thank Robert Talos of Interactive Image Technologies for his technical assistance in the development of these experiments. I especially appreciate the valuable suggestions of my copyeditor, Ben Shriver, and the dedication and talent of the editorial staff of Prentice Hall.

R.H. Berube

Contents

Introduction

Electronics Workbench is similar to a workbench in a real laboratory environment, except that circuits are simulated on a computer and results are obtained more quickly. Electronics Workbench provides all of the components and instruments necessary to create and simulate mixed mode analog and digital circuits on the computer screen. Using a mouse and Electronics Workbench's click and drag schematic editing, you can build a circuit in the central workspace, attach simulated test instruments, simulate actual circuit performance, and display the results on the test instruments. Because circuit faults can be introduced without destroying or damaging actual components, more extensive troubleshooting experiments can be performed using Electronics Workbench. Also, faulty components that are deliberately introduced in a circuit simulated on a computer can help make it easier to find the faulty component in an actual circuit.

Each experiment includes a list of Objectives, a Materials list, a Preparation section, circuit diagrams, and a Procedure section. The Procedure section requires you to record measured data, draw logic circuits, write logic equations, calculate expected values, and answer a series of questions designed to reinforce the theory. The Preparation section provides all of the theory and data needed to complete the procedure. **The Materials list and circuit diagrams make it possible to use this manual in a hardwired laboratory environment.**

If this manual is used in a hardwired laboratory, wire the circuit from the circuit diagram and connect the instruments specified. The pin diagrams for the IC chips required for each experiment are shown in Appendix A. Don't forget to connect each IC chip V_{cc} terminal to +5 V and each GND terminal to the power supply ground. After the digital circuit is wired and checked, turn on the power, record the data in the space provided, and answer the questions.

The CD-ROM provided with this manual has all of the troubleshooting circuits and all of the digital circuits needed to perform the experiments on Electronics Workbench. The CD-ROM is write protected; therefore, you will not be able to save circuit changes. (See Appendix B, Note 4 on how to save circuit changes). You also can wire the circuits on the computer screen yourself using the actual IC chips by selecting the IC chips from the IC chips parts bin. By wiring the actual IC chips in the Electronics Workbench workspace, you can create a hardwired laboratory environment.

In order to perform the experiments in this manual on a computer, you need to install the Electronics Workbench educational version (4.1c or higher) onto your computer system. If Electronics Workbench is not available, it can be obtained from INTERACTIVE IMAGE TECHNOLOGIES, LTD., 111 Peter Street, Suit 801, Toronto, Ontario, Canada M5V2H1. (Tel 1-800-263-5552, Fax 416-977-1818). **The manual provided with the Electronics Workbench software will describe how to use Electronics Workbench**. Additional notes on using Electronics Workbench have been included in Appendix B at the end of this manual.

PART

Combinational Logic Circuits

The experiments in Part I involve the study of logic gates and combinational logic circuits. In combinational logic circuits, the output response follows changes in the input with minimum delay. Prior input conditions have no effect on the present output levels because combinational logic circuits do not have memory.

The logic gates you will study are INVERTERS, OR gates, AND gates, NAND gates, and NOR gates. You will review the Boolean theorems and use them to analyze and simplify combinational logic circuits. You will learn how to use Karnaugh maps to help simplify and design combinational logic circuits. In the final experiment in Part I you will solve some troubleshooting problems in combinational logic circuits.

The logic circuits for the experiments in Part I can be found on the enclosed disk in the PART1 subdirectory.

Name ____________________
Date ____________________

Preliminary Concepts

Objectives:

1. Investigate the Electronics Workbench digital word generator.
2. Investigate the Electronics Workbench logic analyzer.
3. Learn how to use a logic probe to determine logic levels.
4. Determine the voltage levels that represent a binary "one" and a binary "zero."

Materials:

One digital word generator (Electronics Workbench only)
One logic analyzer
One 5 V dc voltage supply
One dc voltmeter
One logic probe
One logic switch
Resistors—500 Ω, 1 kΩ, 2 kΩ, 10 kΩ

Preparation:

Word Generator (Version 4)

The Electronics Workbench Version 4 word generator, shown in Figure 1-1a, can store up to sixteen 8-bit binary words. When the word generator is activated, each of these 8-bit words is transmitted in parallel at its eight output terminals at the bottom of the instrument. To change each bit of a word stored in the word generator, click the bit position with the left mouse button and type "1" or "0." Once a bit has been selected with the mouse, the arrow keys can be used to select a new bit or word.

The 8-bit binary words can be transmitted to the output terminals by clicking the STEP, BURST, or CYCLE buttons on the word generator. To transmit one word at a time, click the STEP button. To transmit all sixteen words in sequence, click the BURST button. To transmit a continuous stream of words, click the CYCLE button. The continuous stream can be stopped by clicking the On-Off switch. The word being transmitted is highlighted. You can select the first word to be transmitted by highlighting the previous word. You can highlight a word by clicking the word number (0–15) to the left of the word.

Each word is transmitted for the duration of one clock period (T) of the internal clock when INTERNAL trigger is selected. The clock frequency (f) can be changed by changing the frequency selection on the front of the word generator. The clock pulses can be monitored at the word generator CLK output terminal. When EXTERNAL trigger is selected, the clock pulse input at the external trigger terminal is used as the clock. The word generator can be triggered on the ascending or descending edge of a clock pulse by selecting the appropriate trigger selection.

For more detailed information about the Version 4 word generator, see your Electronics Workbench Version 4 manual; or you can select the word generator by clicking it with the left mouse button, and then select HELP to bring down the HELP screen.

Word Generator (Version 5)

The Electronics Workbench Version 5 word generator, shown in Figure 1-1b, can store 1024 16-bit binary words. When the word generator is activated, each of these 16-bit words is transmitted in parallel at its sixteen output terminals at the bottom of the instrument. The 16-bit words are stored as 4-character hexadecimal numbers in the left column scroll box. To change a binary word stored in the word generator, change the 4-character hexadecimal number in the scroll box, or type a 16-bit number in the box labeled BINARY (This will change the selected hexadecimal number in the left column scroll box), or type the equivalent ASCII character in the box labeled ASCII. You can select the left column scroll box, the BINARY box, or the ASCII box using the mouse arrow and clicking the left mouse button. When a word in the left column scroll box is selected, its address appears in the EDIT box.

The 16-bit binary words can be transmitted to the output terminals by clicking the STEP, BURST, or CYCLE buttons on the word generator. To transmit one word at a time, click the STEP button. To transmit a group of words in sequence, click the BURST button. To transmit a continuous stream of words, click the CYCLE button. The continuous stream can be stopped by clicking the On-Off switch. The word being transmitted is highlighted in the left column scroll box. You can select the first word to be transmitted by typing the hexadecimal word address in the box labeled INITIAL. You can select the last word to be transmitted by typing the hexadecimal word address in the box labeled FINAL. As the word generator outputs words, each hexadecimal word address appears in the box labeled CURRENT.

Each word is transmitted for the duration of one clock period (T) of the internal clock when INTERNAL trigger is selected. The clock frequency (f) can be changed by changing the frequency selection on the front of the word generator. When EXTERNAL trigger is selected, the clock pulse input at the external trigger terminal is used as the clock. The word generator can be triggered on the ascending or descending edge of a clock pulse by selecting the appropriate trigger selection.

The Version 5 word generator has a breakpoint feature that allows you to set breakpoints. Use a breakpoint when you want to pause a stream of words at a specific word. To insert a breakpoint, select a word in the left scroll window, and then click the BREAKPOINT box on the word generator. To remove a breakpoint, click on an existing breakpoint in the scroll window, and then click the BREAKPOINT box. More than one breakpoint can be set. Breakpoints can be used in both the burst and cycle modes.

For more detailed information about the Version 5 word generator, see your Electronics Workbench Version 5 manual; or you can select the word generator by clicking it with the left mouse button, and then select HELP to bring down the HELP screen.

Logic Analyzer Theory

A logic analyzer is used for acquisition and display of logic states for timing analyis of digital systems. Although a logic analyzer is similar to a multichannel oscilloscope, it displays only multichannel digital data. It samples the state of several channels of rapidly changing digital logic levels on the selected transition of each clock pulse. These logic levels are stored in a memory section. The clock pulses can be generated internally or externally. The clock frequency should be much higher than the frequency of the digital data so that data will not be missed.

A trigger event determines the input data displayed. A logic analyzer can be triggered on the selected transition of an input pulse, the selected transition of the signal applied to the external trigger input, or a particular input binary word. A logic analyzer stores data samples until it reaches the number of samples determined by the pre-trigger setting. It then begins discarding old samples as new samples appear, until it sees the trigger event. After the trigger event, samples are stored until they reach the number of samples determined by the post-trigger setting.

After the trigger event, the stored logic levels are continuously displayed on a CRT screen as multitrace square waves with the trigger point marked. Because of their storage capability, logic analyzers display digital data before and after the trigger event. The pre-trigger multitrace data has already occurred and is not real-time like the waveshapes displayed on an oscilloscope screen.

Logic Analyzer (Version 4)

The Electronics Workbench Version 4 logic analyzer, shown in Figure 1-1a, displays up to eight digital waveshapes on the screen. These eight digital waveshapes are applied at the eight input terminals on the bottom of the instrument. Each of the input terminals corresponds to one horizontal display on the logic analyzer screen. The top row of the display (Channel 0) is the signal entering the first terminal on the left, and the bottom row of the display (Channel 7) is the signal entering the last terminal on the right. The top row displays the most significant bit in a digital word, and the bottom row displays the least significant bit. You can clear the display by clicking the CLEAR button.

The logic analyzer display can be triggered on the ascending or descending edge of an input waveshape, the ascending or descending edge of an external signal, or a specified word pattern. When BURST is selected on the logic analyzer, pulse waveshapes received at the input terminals are used for triggering the display. When EXTERNAL is selected on the logic analyzer, pulse waveshapes received at the external trigger terminal are used for triggering the display. When PATTERN is selected and a bit pattern is stored in the box below the PATTERN button, the stored bit pattern will trigger the display. You can specify a triggering bit pattern by clicking each bit in the box below the PATTERN button and typing a "1" or "0" in the selected bit position. An "x" specifies that either a "1" or a "0" is acceptable.

For more detailed information about the Version 4 logic analyzer, see your Electronics Workbench Version 4 manual; or you can select the logic analyzer by clicking it with the left mouse button, and then select HELP to bring down the HELP screen.

Logic Analyzer (Version 5)

Pull down the File menu and open FIG1-1 on the Version 5 circuits disk to view the Electronics Workbench Version 5 logic analyzer. The Version 5 logic analyzer displays up to sixteen digital waveshapes. These sixteen digital waveshapes are applied at the sixteen input terminals on the left side of the instrument. Each of the input terminals corresponds to one horizontal display on the logic analyzer screen. The top row of the display (Channel 0) is the waveshape entering the top terminal on the left, and the bottom row of the display (Channel 15) is the waveshape entering the bottom terminal on the left. The bottom row displays the most significant bit in a digital word, and the top row displays the least significant bit. If the digital display goes beyond the screen on the right, use the scroll bar on the bottom of the logic analyzer screen to view the remaining waveshapes. You can clear the display by clicking the RESET button.

When the circuit is activated, the logic analyzer samples the input values on the sixteen input terminals and stores the binary data samples until it reaches the pre-trigger number of samples. Then it begins discarding old samples as new samples appear until it sees the trigger signal. After the trigger signal, samples are stored up to the post-trigger number of samples. When the trigger signal is seen, the logic analyzer displays the pre- and post-trigger data on the screen. To dump stored data to the screen when the logic analyzer is not triggered, click the STOP button. If the logic analyzer is already triggered and displaying data, STOP has no effect.

To specify the number of samples stored before and after triggering, click the SET button in the Clock box on the logic analyzer. The Clock Setup dialog box will appear and display the pre-trigger and post-trigger samples. These values can be changed in the dialog box by selecting them with the mouse arrow and clicking the left mouse button.

The logic analyzer can be made to trigger upon reading a specified word or combination of words. To specify the trigger word or word combination, click the SET button in the Trigger box on the logic analyzer. The Trigger Patterns dialog box will appear. You can specify up to three trigger words or word combinations. An x specifies a don't care value. A pattern of all x's will cause the logic analyzer to trigger on the first binary word input.

The clock informs the logic analyzer when to read input samples. The samples can be read on the positive edge or the negative edge of the clock pulses. The clock frequency (Internal clock rate) determines the sampling rate. If the sampling rate is lower than the input data rate, data will be lost. If the sampling rate is higher than the input data rate, the pre-trigger and post-trigger samples settings must be high enough to display all of the input data because each input data word will be sampled more than once. For example, if the clock frequency is ten times the frequency of the input data, each input data word will be sampled ten times. When the INTERNAL clock mode is selected, the internal clock is in control. When the EXTERNAL clock mode is selected, an external clock is in control. To adjust the clock settings, click SET in the Clock box on the logic analyzer and the Clock Setup dialog box will appear. The clock qualifier is an input signal that filters the clock signal. If it is set to x, then the qualifier is disabled and the clock determines when samples are read. If it is set to 1 or 0, the samples are read only when the clock level matches the selected qualifier level.

The Threshold Voltage setting determines the minimum voltage that will be recognized as a binary one. The threshold voltage can be changed by clicking the SET button in the Clock box on the logic analyzer and changing the threshold voltage value in the Clock Setup dialog box.

The red and blue cursor lines on the left and right sides of the logic analyzer screen are used for measuring time differentials between points on the curve plots. These cursor lines can be moved by clicking and dragging them with the mouse arrow in the same way that they are moved on the oscilloscope. The cursor time points can be read in the box at the bottom of the logic analyzer.

For more detailed information about the Version 5 logic analyzer, see your Electronics Workbench Version 5 manual; or you can select the logic analyzer by clicking it with the left mouse button, and then select HELP to bring down the HELP screen.

Logic Voltage Levels

TTL logic gates recognize only high and low input voltages and produce only high and low output voltages. A voltage between 2 V and 5 V is recognized as a logical high (1) and a voltage between 0 V and 0.8 V is recognized as a logical low (0) by TTL logic device inputs. Voltages between 0.8 V and 2 V cannot be interpreted by TTL logic devices.

A logic probe is used to detect a binary high (1) or a binary low (0) at various terminals in a digital logic circuit. In the Electronics Workbench logic probe, a light is displayed when a binary high (1) is present (voltage above 2 V) and the light will go out when a binary low (0) is present (voltage below 0.8 V). A hardwired logic probe will normally have three lights, one for detecting a binary high, one for detecting a binary low, and one for detecting an open circuit or a voltage that is out-of-range.

A logic switch is used to input a binary high (2–5 V) or a binary low (0–0.8 V) to a logic circuit, as shown in Figure 1-2. The circuit in Figure 1-2 will demonstrate how logic voltage levels are measured using a logic probe and how a logic voltage level can be changed using a logic switch.

Figure 1-1a Word Generator and Logic Analyzer (EWB Version 4)

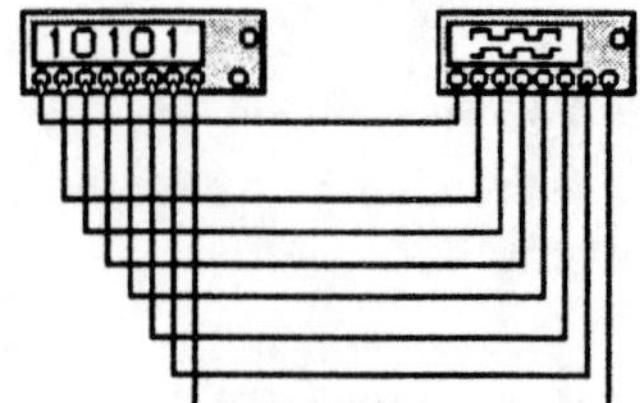

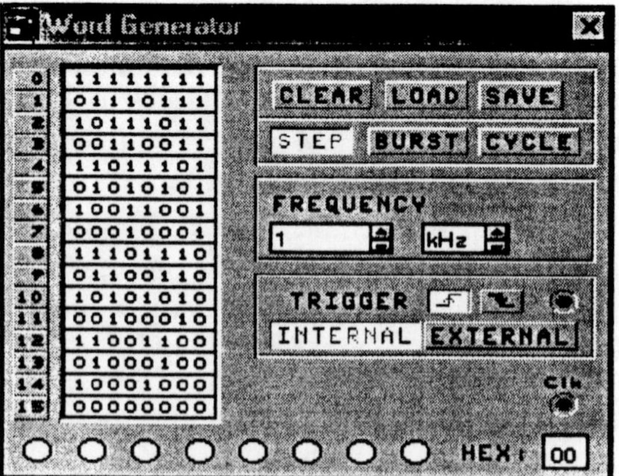

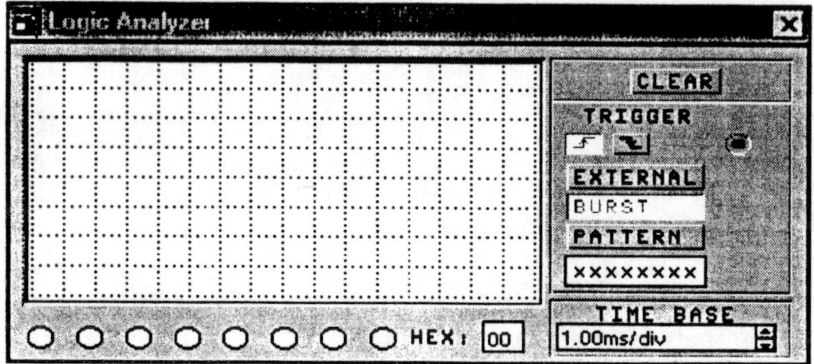

Figure 1-1b Word Generator and Logic Analyzer (EWB Version 5)

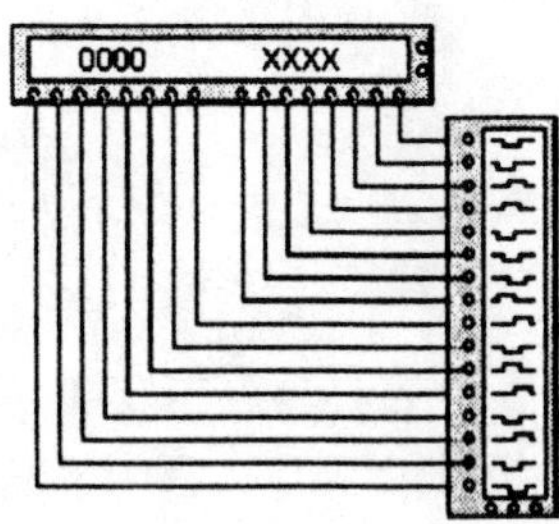

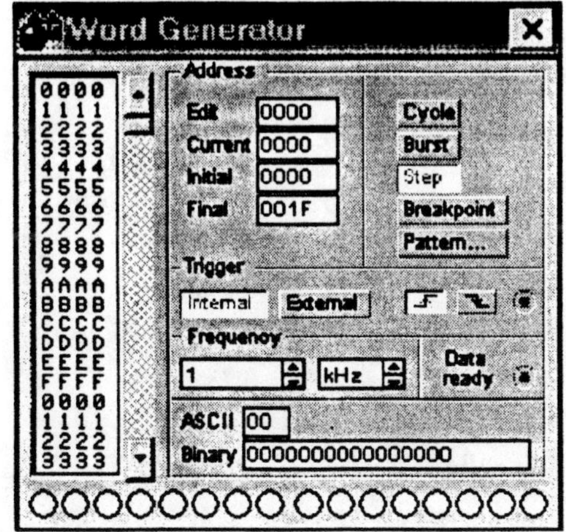

Logic Analyzer Settings
Clocks per division ---------------- 8

Clock Setup dialog box
Clock edge --------------- positive
Clock mode -------------- Internal
Internal clock rate ------ 10 kHz
Clock qualifier ----------- x
Pre-trigger samples --- 160
Post-trigger samples -- 160
Threshold voltage (V) 2

Trigger Patterns dialog box
A ----------------------------- xxxxxxxxxxxxxxxx
Trigger combinations -- A
Trigger qualifier ---------- x

Figure 1-2 Logic Voltage Levels

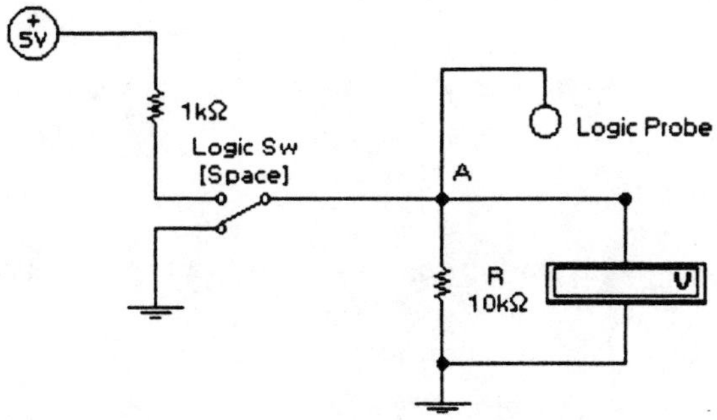

NOTE: If you are using this manual in a hardwired laboratory, see your lab instructor for information on the instruments and equipment you will be using. Because a word generator may not be available, you may need to start this experiment at the Logic Voltage Levels section of the procedure.

Procedure:

Word Generator and Logic Analyzer (Version 4)

If you are using Electronics Workbench Version 5, skip this section and go to the Word Generator and Logic Analyzer (Version 5) section.

1. Pull down the File menu and open FIG1-1. Make sure that the word generator and logic analyzer settings are as shown in Figure 1-1a. Click STEP on the word generator to start the analysis. Notice that the first 8-bit binary word (Word 0) has been transmitted and displayed on the logic analyzer screen from top-to-bottom.

Question: Is the 8-bit binary word transmitted by the word generator (Word 0) the same 8-bit word displayed on the logic analyzer screen from top-to-bottom? What is the binary word?

2. Click STEP on the word generator again and notice that the second 8-bit binary word (Word 1) has been transmitted and displayed on the logic analyzer screen from top-to-bottom

Question: Is the second 8-bit binary word transmitted by the word generator (Word 1) the same 8-bit word displayed on the logic analyzer screen from top-to-bottom? What is the binary word?

3. Continue to click STEP on the word generator until all sixteen of the 8-bit binary words (Words 0–15) have been transmitted and displayed on the logic analyzer screen.

Questions: Are all sixteen of the transmitted binary words the same words displayed on the logic analyzer screen from top-to-bottom?

Why is each binary word displayed from top-to-bottom on the logic analyzer screen?

4. Click BURST on the word generator. Notice that the word generator transmitted all sixteen binary words (Words 0–15) in sequence, starting with the word following the current word selected.

Question: Do the binary words transmitted by the word generator still match the words displayed on the logic analyzer screen from top-to-bottom?

5. Click CYCLE on the word generator. Notice that the word generator continues to cycle through all sixteen binary words.

6. Click the On-Off switch to stop the analysis. Click the number 15 on the word generator to reset the generator to the last binary word. Change the frequency selection on the word generator to 2 kHz. Click CYCLE to start the analysis run.

Question: Did the pulse time period change for each of the pulses displayed on the logic analyzer screen? Why?

7. Click the On-Off switch to stop the analysis run. Click the number 15 on the word generator to reset the generator to the last binary word. Return the frequency selection on the word generator to 1 kHz. Set the word pattern on the *logic analyzer* to the pattern of Word 6 in the *word generator*, and then click PATTERN on the *logic analyzer*. Click BURST on the *word generator*. Notice that the logic analyzer did not display the word patterns of Words 0–5 and started the display with Word 6.

8. Click the On-Off switch to stop the analysis run. Click BURST on the *logic analyzer*. Remove the black wire from the first terminal on the word generator and connect it to the CLK terminal. Click CYCLE on the word generator to start the analysis run. Notice that a 1 kHz clock pulse (T = 1 msec) has been displayed on the top line of the logic analyzer (black curve plot on Channel 0).

9. Change the frequency selection on the word generator to 2 kHz.

Question: What happened to the clock pulse (black curve plot on Channel 0) on the logic analyzer screen? Explain.

10. Change the TIME BASE on the logic analyzer to 0.5 ms/div.

Question: What happened to the clock pulse (black curve plot on Channel 0) on the logic analyzer screen? Explain.

Word Generator and Logic Analyzer (Version 5)

1. Pull down the File menu and open FIG1-1. Make sure that the word generator and logic analyzer settings are as shown in Figure 1-1b. Click STEP on the word generator to start the analysis. Notice that the first 16-bit binary word has been transmitted and displayed on the logic analyzer screen from bottom-to-top.

Question: Is the 16-bit binary word transmitted by the word generator the same 16-bit word displayed on the logic analyzer screen from bottom-to-top? What is the binary word?

2. Click STEP on the word generator again and notice that the second 16-bit binary word has been transmitted and displayed on the logic analyzer screen from bottom-to-top.

Question: Is the second 16-bit binary word transmitted by the word generator the same 16-bit word displayed on the logic analyzer screen from bottom-to-top? What is the binary word?

3. Continue to click STEP on the word generator until sixteen of the 16-bit binary words have been transmitted and displayed on the logic analyzer screen.

Questions: Are all sixteen of the transmitted binary words the same words displayed on the logic analyzer screen from bottom-to-top?

Why is each binary word displayed from bottom-to-top on the logic analyzer screen?

4. Click the On-Off switch to stop the analysis run. Click BURST on the word generator. Notice that the word generator transmitted 32 (001FH) binary words in sequence, but the logic analyzer only displayed 16 binary words. Note the Internal clock rate and the Post-trigger samples setting on the logic analyzer.

Question: Why were only 16 binary words displayed on the logic analyzer screen even though 32 binary words were transmitted by the word generator?

5. Click the On-Off switch to stop the analysis run. Click the SET button in the Clock box on the logic analyzer. Change the Post-trigger samples to 320 and click ACCEPT. Change CLOCKS PER DIVISION to 16 on the logic analyzer. Click BURST on the word generator.

Question: How many binary words were displayed on the logic analyzer screen? Did it change from the number in Step 4? Explain.

6. Click the On-Off switch to stop the analysis run. Click CYCLE on the word generator. Notice that the word generator cycles through binary words continuously.

Question: What happened on the logic analyzer screen? Explain.

7. Click the On-Off switch to stop the analysis run. Click the SET button in the Clock box on the logic analyzer. Change Pre-trigger samples to 60 and Post-trigger samples to 260 in the Clock Setup dialog box, and then click ACCEPT. Click the SET button in the Trigger box on the logic analyzer. Change the A word to 0110011001100110 (6666H), select A for Trigger combinations, and then click ACCEPT. Click CYCLE on the *word generator*. Notice that the logic analyzer did not display the word patterns until the trigger signal 0110011001100110 (6666H), then it displayed the pre-trigger (before the line) and post-trigger (after the line) word patterns.

Questions: How many pre-trigger 16-bit binary words were displayed?

How many post-trigger 16-bit binary words were displayed?

Did this satisfy the logic analyzer settings?

8. Click the On-Off switch to stop the analysis run. Click the SET button in the Clock box on the logic analyzer. Change the Pre-trigger samples setting to 40 and the Post-trigger samples setting to 100, and then click ACCEPT. Click CYCLE on the word generator to start the analysis run.

Questions: How many pre-trigger 16-bit binary words were displayed?

How many post-trigger 16-bit binary words were displayed?

Explain any difference from the Step 7 results.

9. Click the On-Off switch to stop the analysis run. Click the SET button in the Clock box on the logic analyzer. Change the clock frequency (Internal clock rate) to 20 kHz, and then click ACCEPT. Click CYCLE on the word generator to start the analysis run.

Questions: How many pre-trigger 16-bit binary words were displayed?

How many post-trigger 16-bit binary words were displayed?

Explain any difference from the results in Step 8.

Logic Voltage Levels

1. Click the On-Off switch to stop the analysis. Pull down the File menu and open FIG1-2. Click the On-Off switch to run the analysis. Record the voltage on terminal A (V_A).

V_A = ______________

Questions: Does voltage V_A represent a binary "zero" or a binary "one?"

Is the logic probe light on or off? Does this indicate a binary "zero" or a binary "one?"

2. Press the space bar on the computer keyboard to change the logic switch to the 5 V dc supply voltage. Record the voltage on terminal A (V_A).

V_A = ______________

Questions: Does voltage V_A represent a binary "zero" or a binary "one?"

Is the logic probe light on or off? Does this indicate a binary "zero" or a binary "one?"

Why is the voltage less than 5 V?

3. Click the On-Off switch to stop the analysis. Change Resistor R to 2 kΩ. Click the On-Off switch to run the analysis again. Record the voltage at terminal A (V_A).

V_A = ______________

Questions: Was voltage V_A the same as in Step 2? If not, why not?

Does this represent a binary "zero" or a binary "one?" Does the logic probe verify this?

4. Click the On-Off switch to stop the analysis. Change Resistor R to 500 Ω. Click the On-Off switch to run the analysis again. Record the voltage at terminal A (V_A).

V_A = ______________

Questions: Is the logic probe light still on? If not, why not?

Does voltage V_A represent a binary "zero" or a binary "one?" Explain.

Name ________________________________
Date ________________________________

Logic Gates: INVERTER, OR, and AND

Objectives:

1. Investigate the behavior of the INVERTER logic gate.
2. Investigate the behavior of a two-input OR logic gate.
3. Investigate the behavior of a two-input AND logic gate.
4. Determine the effect of loading on a digital logic gate output.

Materials:

One 5 V dc voltage supply
Two logic switches
Three dc voltmeters
Three logic probe lights
One logic analyzer
Two pulse generators
One INVERTER (1-7404 IC)
One two-input OR gate (1-7432 IC)
One two-input AND gate (1-7408 IC)
Resistors—10 Ω, 100 Ω, 1 kΩ, and 10 kΩ

Preparation:

The INVERTER logic gate inverts the input. Therefore, when the input is low (0) the output will be high (1) and when the input is high (1) the output will be low (0). Circuits for studying the INVERTER logic gate are shown in Figures 2-1 and 2-2.

The OR gate output will be high (1) when one or more inputs are high (1). The OR gate output will be low (0) only when all of the inputs are low (0). Circuits for studying the two-input OR logic gate are shown in Figures 2-3 and 2-4.

The AND gate output will be high (1) only when all of the inputs are high (1). If any of the AND gate inputs are low (0), then the output will be low (0). Circuits for studying the two-input AND logic gate are shown in Figures 2-5 and 2-6.

If the load resistance on any of the TTL logic gates is too low and draws too much current from the gate output terminal, the output voltage will be dragged down below 2 V when the output is trying to produce a high (1) output. This will produce an error in the logical network output because other TTL logic gate inputs will not recognize an input below 2 V as a logical high (1). Fan out for any TTL logic gate is the number of TTL logic gate inputs that can be connected to a TTL logic gate output before the above problem of overloading will occur.

Figure 2-1 The INVERTER Gate

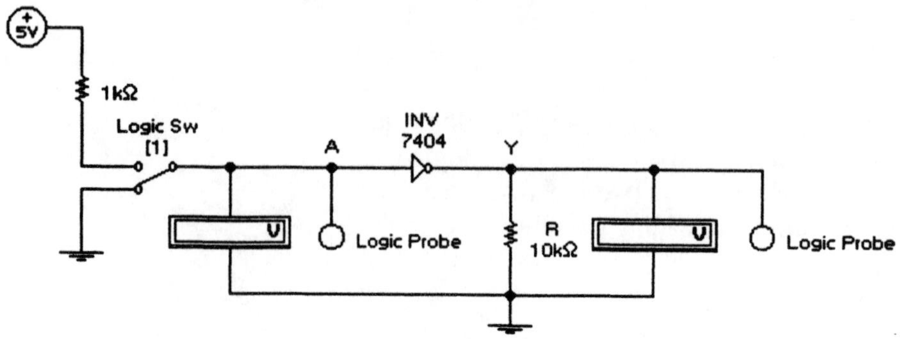

Figure 2-2a INVERTER Pulse Reponse (EWB Version 4)

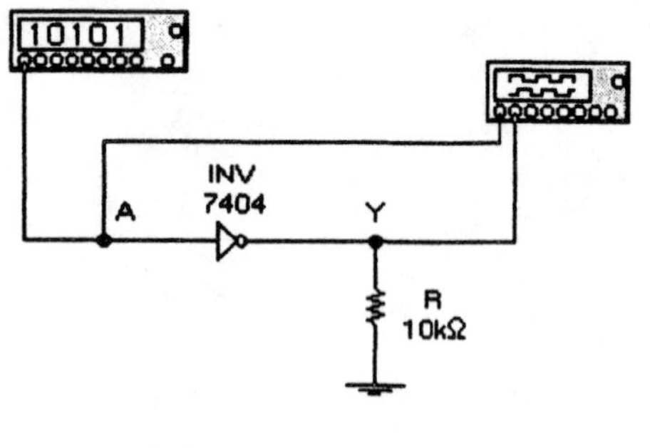

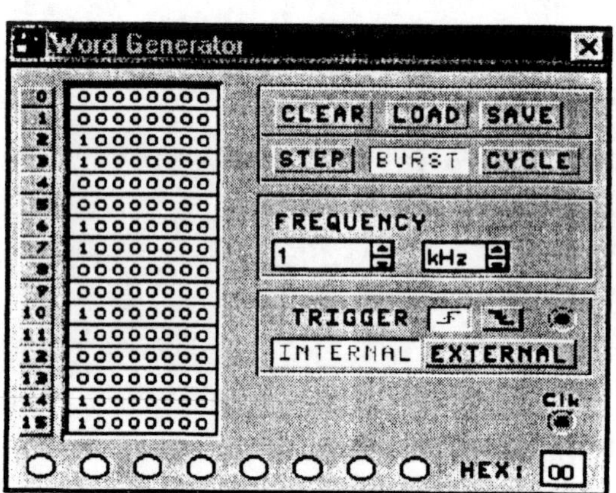

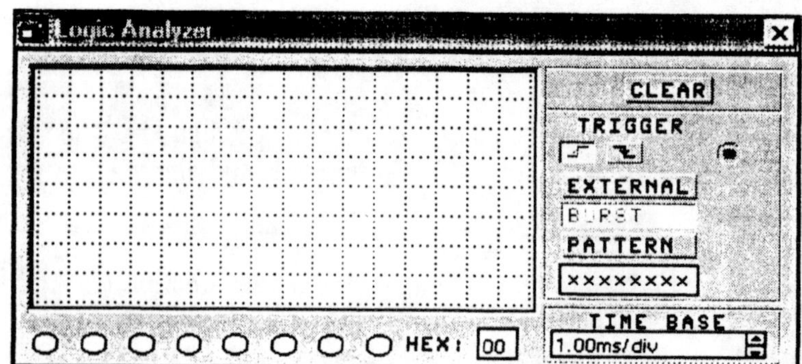

Figure 2-2b INVERTER Pulse Response (EWB Version 5)

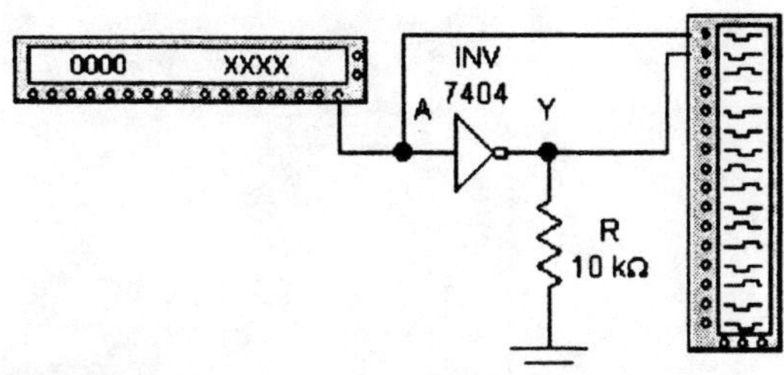

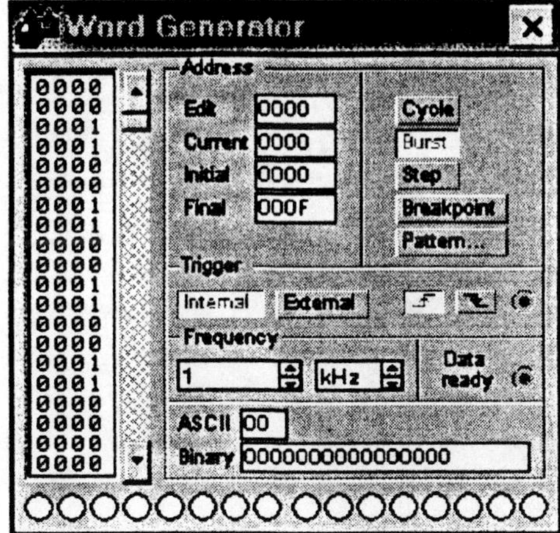

Logic Analyzer Settings
Clocks per division ---------------- 8

Clock Setup dialog box
Clock edge --------------- positive
Clock mode -------------- Internal
Internal clock rate ------ 10 kHz
Clock qualifier ----------- x
Pre-trigger samples --- 100
Post-trigger samples -- 1000
Threshold voltage (V) 2

Trigger Patterns dialog box
A ----------------------------- xxxxxxxxxxxxxxxx
Trigger combinations -- A
Trigger qualifier ---------- x

Figure 2-3 The OR Gate

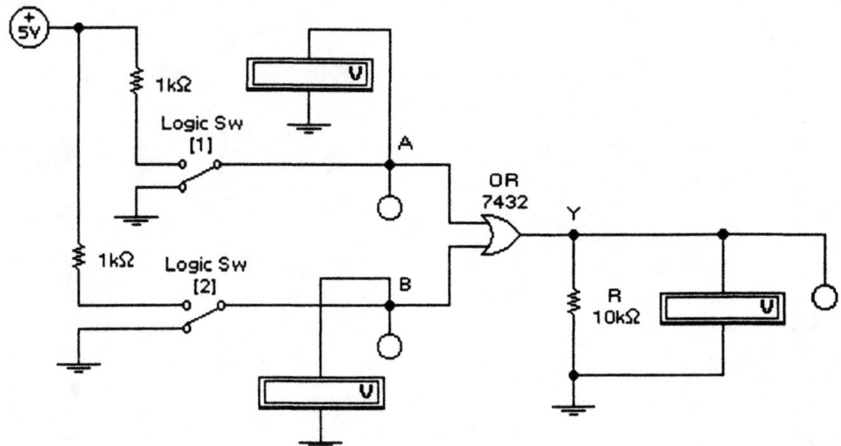

Figure 2-4a OR Gate Pulse Response (EWB Version 4)

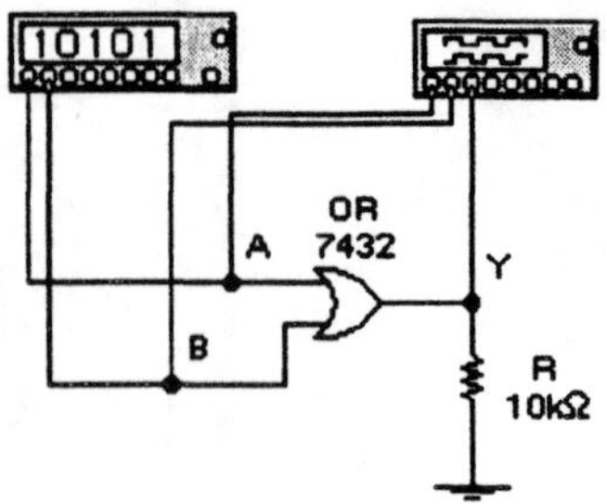

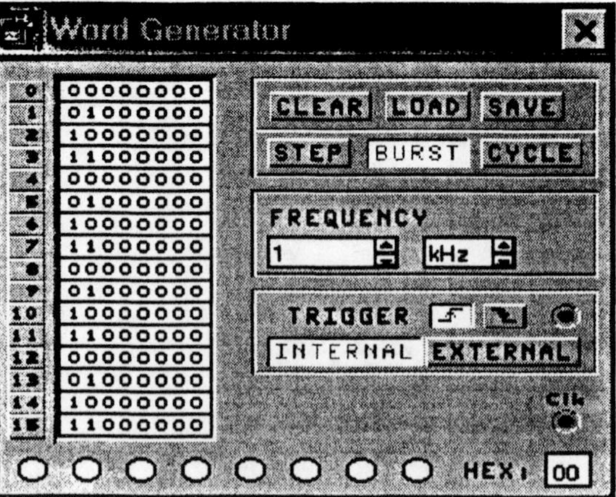

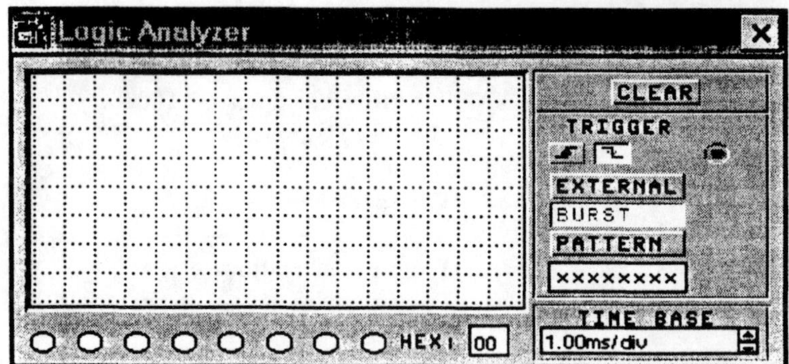

Figure 2-4b OR Gate Pulse Response (EWB Version 5)

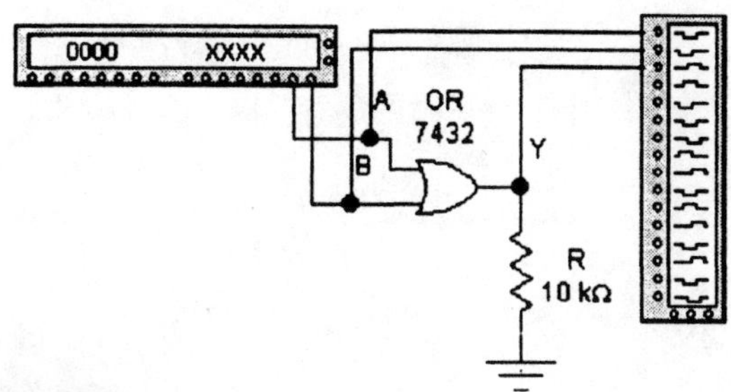

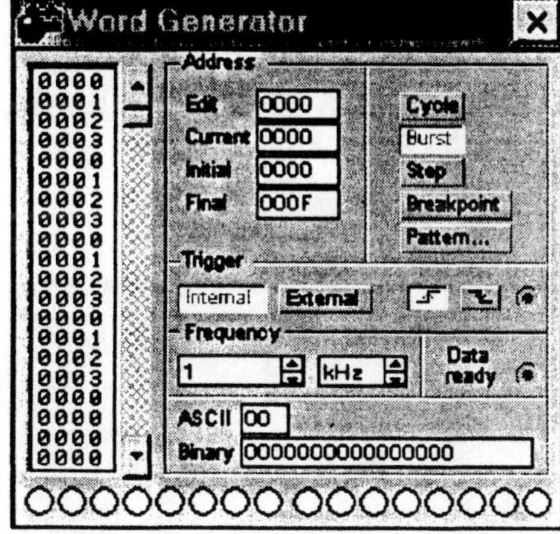

Logic Analyzer Settings
Clocks per division ---------------- 8

Clock Setup dialog box
Clock edge -------------- positive
Clock mode -------------- Internal
Internal clock rate ------ 10 kHz
Clock qualifier ----------- x
Pre-trigger samples --- 100
Post-trigger samples -- 1000
Threshold voltage (V) 2

Trigger Patterns dialog box
A ------------------------------ xxxxxxxxxxxxxxxx
Trigger combinations -- A
Trigger qualifier ---------- x

Figure 2-5 The AND Gate

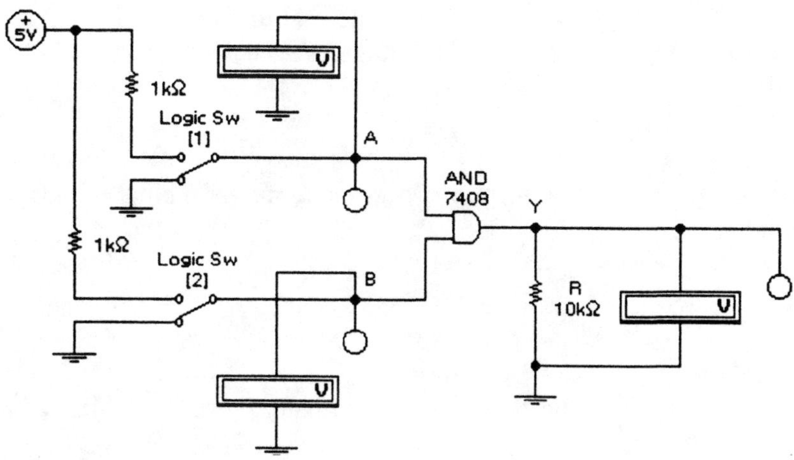

Figure 2-6a AND Gate Pulse Response (EWB Version 4)

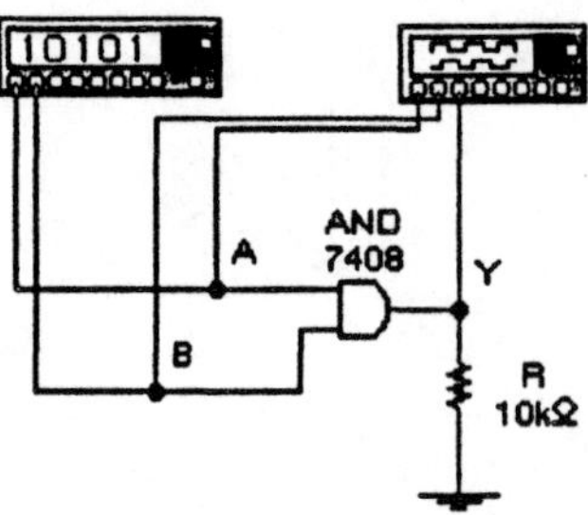

Figure 2-6b AND Gate Pulse Response (EWB Version 5)

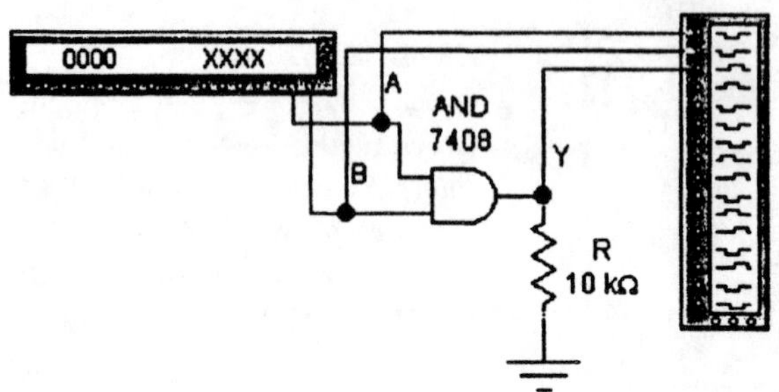

Procedure:

INVERTER

1. Pull down the File menu and open FIG2-1. Notice that the logic switch is placing a zero (ground equals 0 V) on the input (A) of the INVERTER gate. Resistor R represents the load on the INVERTER output. Click the On-Off switch to run the analysis. Record the INVERTER input voltage (V_A) and output voltage (V_Y).

   V_A = ____________ V_Y = ____________

Questions: Is the output probe light on or off? Does this represent a binary "zero" or a binary "one" on the INVERTER output?

Is the input probe light on or off? Does this represent a binary "zero" or a binary "one" on the INVERTER input?

Is the INVERTER output voltage above 2 V? Does this represent a binary "zero" or a binary "one?"

Why is the INVERTER output voltage less than 5 V?

2. Click the On-Off switch to stop the analysis run. Change Resistor R to 100 Ω. Click the On-Off switch to run the analysis again. Record the INVERTER output voltage (V_Y).

V_Y = ____________

Question: How does the new INVERTER output voltage compare with the output voltage in Step 1? Why are they different?

3. Click the On-Off switch to stop the analysis run. Change Resistor R to 10 Ω. Click the On-Off switch to run the analysis again. Record the INVERTER output voltage (V_Y).

V_Y = ____________

Question: Is the INVERTER output voltage above 2 V? If not, why, not?

Is the logic probe light on? If not, why not?

4. Click the On-Off switch to stop the analysis run. Change Resistor R back to 10 kΩ. Press the "1" key on the keyboard to change the logic switch to a binary "one" input (5 V source). Click the On-Off switch to run the analysis again. Record the INVERTER input voltage (V_A) and output voltage (V_Y).

V_A = ____________ V_Y = ____________

Question: Is the output probe light on or off? Does this represent a binary "zero" or a binary "one" on the INVERTER output?

Is the input probe light on or off? Does this represent a binary "zero" or a binary "one" on the INVERTER input?

Is the INVERTER output voltage below 0.8 V? Does this represent a binary "zero" or a binary "one"?

5. Keep pressing the "1" key on the keyboard while the analysis is running to make the INVERTER input voltage switch up and down.

Question: What is the relationship between the binary input and the binary output for the INVERTER gate?

6. Click the On-Off switch to stop the analysis run. Pull down the File menu and open FIG2-2. The word generator and logic analyzer settings should be as shown in Figure 2-2. Click the On-Off switch to run the analysis. Notice that the word generator has applied a pulse pattern to the INVERTER input, shown in red on the logic analyzer screen. The blue curve plot is the INVERTER output. Draw the INVERTER input and output curve plots in the space provided and label them.

> NOTE: In a hardwired laboratory, use a pulse generator to apply a square wave to the INVERTER input and use a dual trace oscilloscope to monitor the INVERTER input and output, if a logic analyzer is not available.

Question: What is the relationship between the INVERTER input and output curve plots? Is it what you expected? Explain.

OR Gate

7. Click the On-Off switch to stop the analysis run. Pull down the File menu and open FIG2-3. Notice that Logic Switches 1 and 2 are placing binary "zeros" (ground equals 0 V) on the two OR gate inputs (A and B). Resistor R represents the load on the OR gate output. Click the On-Off switch to run the analysis. Record the output voltage (V_Y) for the binary inputs (A and B) in Table 2-1.

Table 2-1 OR Gate

A	B	V_Y(V)
0	0	
0	1	
1	0	
1	1	

8. By pressing the "1" and "2" keys on the keyboard to change the logic switch positions, change the binary inputs (A and B) to the remaining values in Table 2-1 and record the output voltage (V_Y) for each case.

Questions: Is the OR gate output voltage above 2 V when a binary "one" is applied to one or more of the inputs? Is this output a binary "one" or a binary "zero"? Do the logic probe lights confirm this?

Is the OR gate output voltage below 0.8 V when a binary "zero" is applied to both gate inputs? Is this output a binary "one" or a binary "zero"? Do the logic probe lights confirm this?

Based on the results in Table 2-1, what conclusion can you draw about the relationship between the OR gate binary output and the binary inputs?

9. **Click the On-Off switch to stop the analysis run. Pull down the File menu and open FIG2-4. The word generator and logic analyzer settings should be as shown in Figure 2-4. Click the On-Off switch to run the analysis. Notice that the word generator has applied a pulse pattern to each OR gate input, shown in red and green on the logic analyzer screen. The blue curve plot is the OR gate output. Draw the OR gate input and output curve plots in the space provided and label them.**

> NOTE: In a hardwired laboratory, use pulse generator outputs to apply square waves to the OR gate inputs.

Question: Does the OR gate output go low only when both inputs are low? Is this expected for an OR gate?

AND Gate

10. Click the On-Off switch to stop the analysis run. Pull down the File menu and open FIG2-5. Notice that Logic Switches 1 and 2 are placing binary zeros (ground equals 0 V) on the two AND gate inputs (A and B). Resistor R represents the load on the AND gate output. Click the On-Off switch to run the analysis. Record the output voltage (V_Y) for the binary inputs (A and B) in Table 2-2.

Table 2-2 AND Gate

A	B	$V_Y(V)$
0	0	
0	1	
1	0	
1	1	

11. By pressing the "1" and "2" keys on the keyboard to change the logic switch positions, change the binary inputs (A and B) to the remaining values in Table 2-2 and record the output voltage (V_Y) for each case.

Questions: Is the AND gate output voltage above 2 V when a binary "one" is applied to both gate inputs? Is this output a binary "one" or a binary "zero"? Do the logic probe lights confirm this?

Is the AND gate output voltage below 0.8 V when a binary "zero" is applied to one or more of the inputs? Is this output a binary "one" or a binary "zero"? Do the logic probe lights confirm this?

Based on the results in Table 2-2, what conclusion can you draw about the relationship between the AND gate binary output and the binary inputs?

12. Click the On-Off switch to stop the analysis run. Pull down the File menu and open FIG2-6. The word generator and logic analyzer settings should be as shown in Figure 2-4. Click the

On-Off switch to run the analysis. Notice that the word generator has applied a pulse pattern to each AND gate input, shown in red and green on the logic analyzer screen. The blue curve plot is the AND gate output. Draw the AND gate input and output curve plots in the space provided and label them.

NOTE: In a hardwired laboratory, use pulse generator outputs to apply square waves to the AND gate inputs.

Question: Does the AND gate output go high only when both inputs are high? Is this expected for an AND gate?

Name ______________________________
Date ______________________________

Logic Gates: NAND and NOR

Objectives:

1. Complete the truth table for an AND gate.
2. Complete the truth table for a NAND gate and compare it with the AND gate truth table.
3. Plot the NAND gate pulse response timing diagram.
4. Complete the truth table for an OR gate.
5. Complete the truth table for a NOR gate and compare it with the OR gate truth table.
6. Plot the NOR gate pulse response timing diagram.

Materials:

One 5 V dc voltage supply
Two logic switches
Three logic probe lights
One logic analyzer
Two pulse generators
One two-input OR gate (1-7432 IC)
One two-input AND gate (1-7408 IC)
One two-input NAND gate (1-7400 IC)
One two-input NOR gate (1-7402 IC)
Resistors—1 kΩ and 10 kΩ

Preparation:

A truth table lists the outputs for all of the possible input combinations for a logic gate or logic network. See the Preparation section in Experiment 2 for information on the OR and AND logic gates.

The NAND gate is equivalent to an AND gate with an INVERTER on the output. Therefore, the NAND gate output is the inverse of the AND gate output. The NAND gate output will be low (0) only when all of the inputs are high (1). The NAND gate output will be high (1) if any of the inputs are low (0). Circuits for plotting the truth table for the two-input AND and NAND logic gates are shown in Figures 3-1 and 3-2, respectively. The circuit for plotting a NAND gate pulse response timing diagram is shown in Figure 3-3.

The NOR gate is equivalent to an OR gate with an INVERTER on the output. Therefore, the NOR gate output is the inverse of the OR gate output. The NOR gate output will be high (1) only when all of the inputs are low (0). The NOR gate output will be low (0) if any of the inputs are high (1). Circuits for plotting the truth table for the two-input OR and NOR logic gates are shown in Figures 3-4 and 3-5, respectively. The circuit for plotting a NOR gate pulse response timing diagram is shown in Figure 3-6.

Figure 3-1 AND Gate

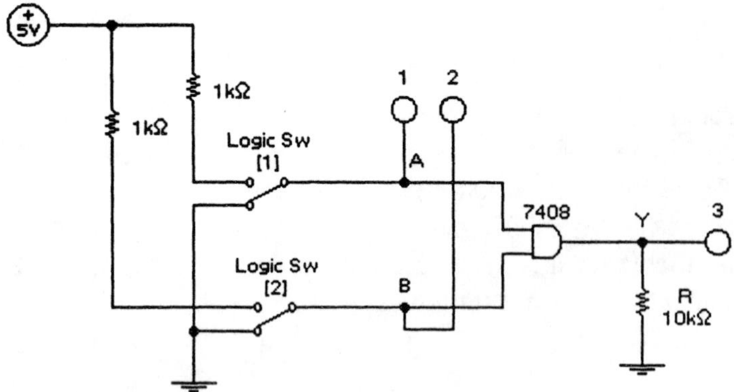

Figure 3-2 NAND Gate

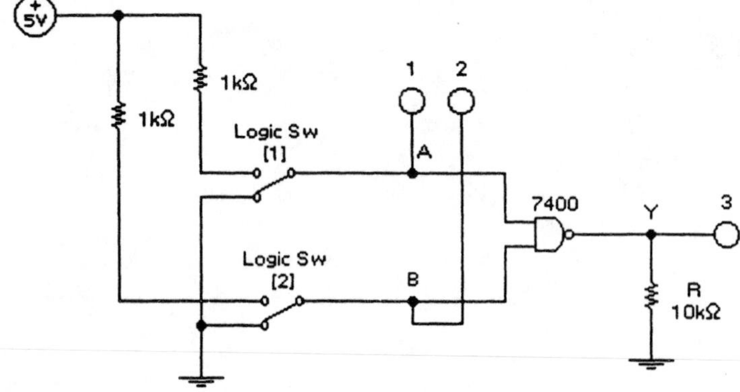

Figure 3-3a NAND Gate Pulse Response (EWB Version 4)

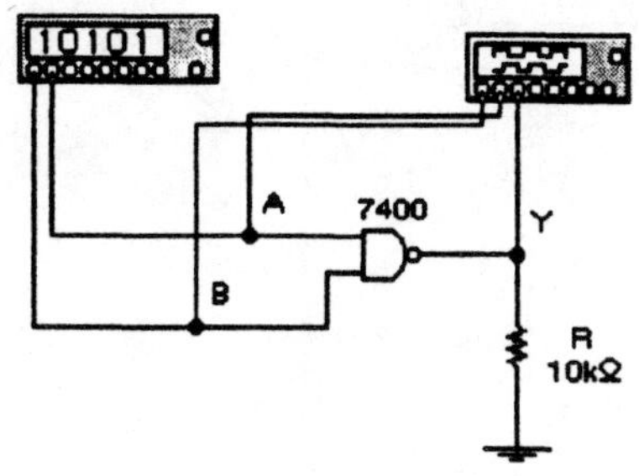

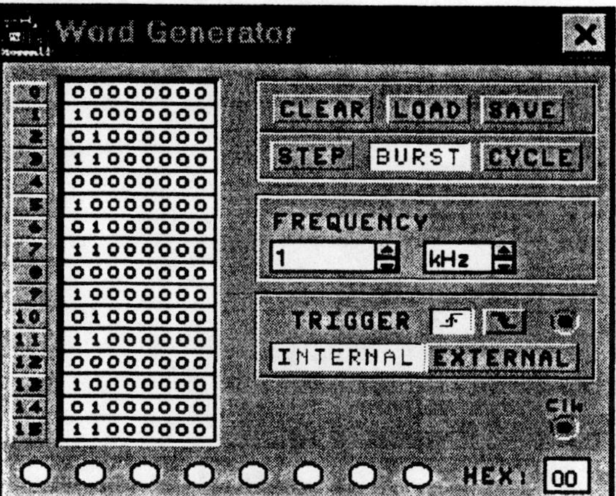

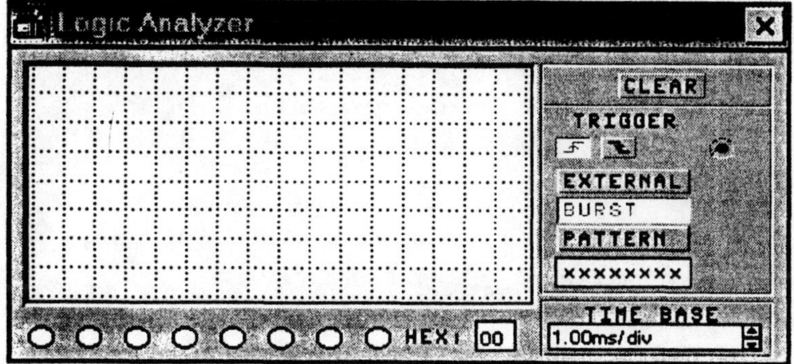

Figure 3-3b NAND Gate Pulse Response (EWB Version 5)

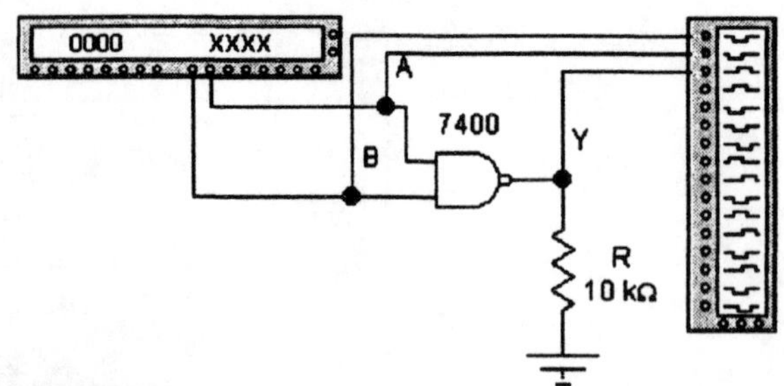

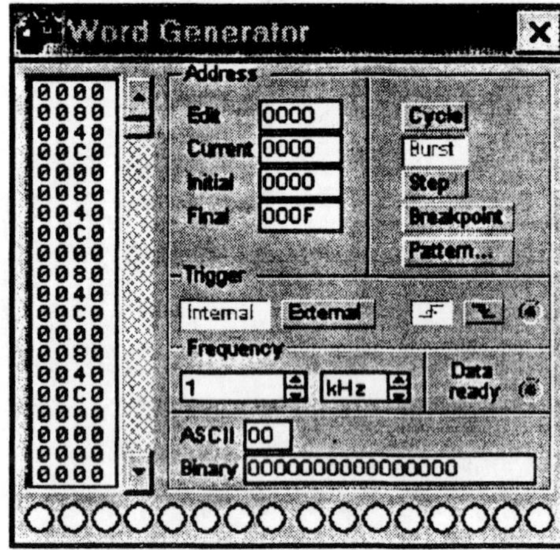

Logic Analyzer Settings
Clocks per division ---------------- 8

Clock Setup dialog box
Clock edge --------------- positive
Clock mode -------------- Internal
Internal clock rate ------ 10 kHz
Clock qualifier ----------- x
Pre-trigger samples --- 100
Post-trigger samples -- 1000
Threshold voltage (V) 2

Trigger Patterns dialog box
A ------------------------------ xxxxxxxxxxxxxxxx
Trigger combinations -- A
Trigger qualifier ---------- x

Figure 3-4 OR Gate

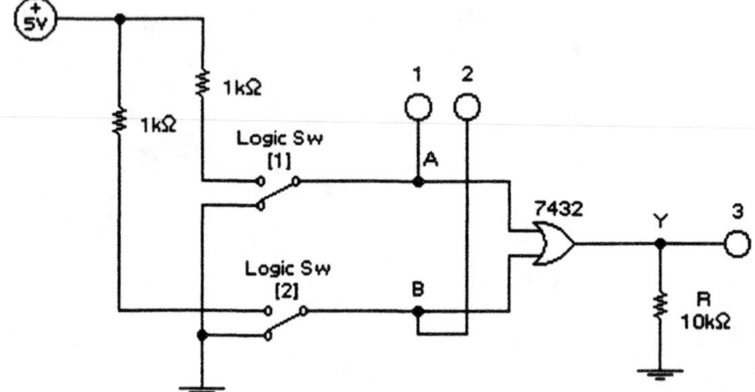

Figure 3-5 NOR Gate

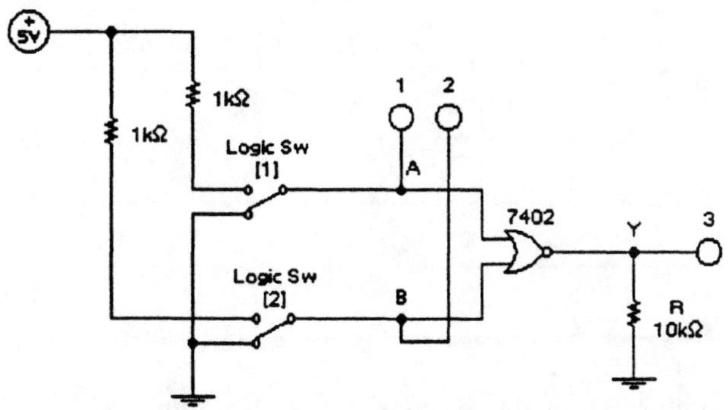

Figure 3-6a NOR Gate Pulse Response (EWB Version 4)

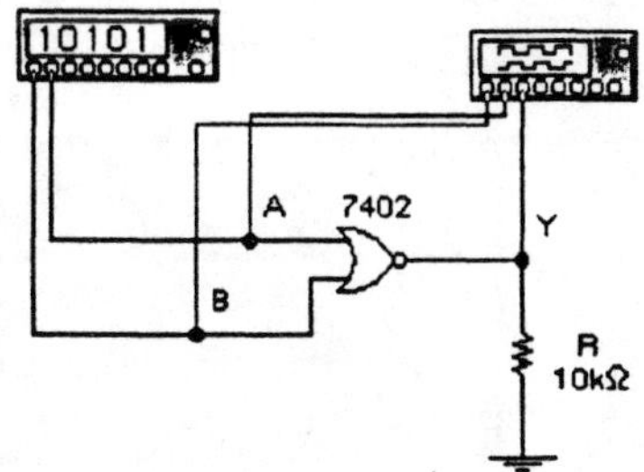

Figure 3-6b NOR Gate Pulse Response (EWB Version 5)

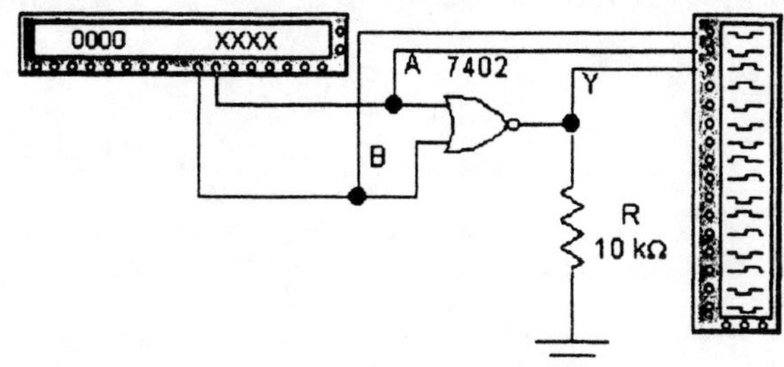

Procedure:

1. Pull down the File menu and open FIG3-1. Notice that Logic Switches 1 and 2 are placing binary "zeros" (ground equals 0 V) on the two AND gate inputs (A and B). Resistor R represents the load on the AND gate output. Click the On-Off switch to run the analysis. By switching the logic switches to the appropriate positions, complete the AND gate truth table (Table 3-1). To switch Logic Switch 1, type "1" on the keyboard. To switch Logic Switch 2, type "2" on the keyboard.

Table 3-1 AND Gate Truth Table

A	B	Y
0	0	
0	1	
1	0	
1	1	

2. Click the On-Off switch to stop the analysis run. Pull down the File menu and open FIG3-2. Click the On-Off switch to run the analysis. By switching the logic switches to the appropriate positions, complete the NAND gate truth table (Table 3-2).

Table 3-2 NAND Gate Truth Table

A	B	Y
0	0	
0	1	
1	0	
1	1	

Question: What is the difference between the AND gate truth table and the NAND gate truth table? Explain.

3. Click the On-Off switch to stop the analysis run. Pull down the File menu and open FIG3-3. The word generator and logic analyzer settings should be as shown in Figure 3-3. Click the On-Off switch to run the analysis. Notice that the word generator has applied a pulse pattern to each NAND gate input, shown in red and green on the logic analyzer screen. The blue curve plot is the NAND gate output. Draw the NAND gate input and output curve plots in the space provided and label them.

> NOTE: In a hardwired laboratory, use pulse generator outputs to apply square waves to the NAND gate inputs.

Question: Does the NAND gate output go low only when all of the inputs are high? Is this expected for a NAND gate?

4. Click the On-Off switch to stop the analysis run. Pull down the File menu and open FIG3-4. Notice that Logic Switches 1 and 2 are placing binary zeros (ground equals 0 V) on the two OR gate inputs (A and B). Resistor R represents the load on the OR gate output. Click the On-Off switch to run the analysis. By switching the logic switches to the appropriate positions, complete the OR gate truth table (Table 3-3).

Table 3-3 OR Gate Truth Table

A	B	Y
0	0	
0	1	
1	0	
1	1	

5. Click the On-Off switch to stop the analysis run. Pull down the File menu and open FIG3-5. Click the On-Off switch to run the analysis. By switching the logic switches to the appropriate positions, complete the NOR gate truth table (Table 3-4).

Table 3-4 NOR Gate Truth Table

A	B	Y
0	0	
0	1	
1	0	
1	1	

Question: What is the difference between the OR gate truth table and the NOR gate truth table? Explain.

6. Click the On-Off switch to stop the analysis run. Pull down the File menu and open FIG3-6. The word generator and logic analyzer settings should be as shown in Figure 3-3. Click the On-Off switch to run the analysis. Notice that the word generator has applied a pulse pattern to each NOR gate input, shown in red and green on the logic analyzer screen. The blue curve plot is the NOR gate output. Draw the NOR gate input and output curve plots in the space provided and label them.

NOTE: In a hardwired laboratory, use pulse generator outputs to apply square waves to the NOR gate inputs.

Question: Does the NOR gate output go high only when all of the inputs are low? Is this expected for a NOR gate?

Name ______________________

Date ______________________

Boolean Theorems

Objectives:

1. Verify experimentally some of the single-variable and multivariable Boolean theorems.
2. Verify experimentally DeMorgan's theorems.

Materials:

One 5 V dc voltage supply
Three logic switches
Five logic probe lights
Four two-input AND gates (1-7408 IC)
Two INVERTERS (1-7404 IC)
Four two-input OR gates (1-7432 IC)
One two-input NAND gate (1-7400 IC)
One two-input NOR gate (1-7402 IC)
One 1 kΩ resistor

Preparation:

Logic gates and logic networks can be described and analyzed using Boolean equations. The Boolean equation for a logic network can be simplified or modified by using one or more of the Boolean theorems. A simplified or modified logic network can then be drawn from the modified Boolean equation. In this experiment, you will verify some of the Boolean theorems that you will be using to simplify and modify logic networks.

Figure 4-1 shows four logic circuits that will be used to verify the following single-variable Boolean theorems involving the AND operation.

$$(X)(0) = 0$$
$$(X)(1) = X$$
$$(X)(X) = X$$
$$(X)(X') = 0$$

Figure 4-2 shows four logic circuits that will be used to verify the following single-variable Boolean theorems involving the OR operation.

X + 0 = X
X + 1 = 1
X + X = X
X + X' = 1

Figures 4-3, 4-4, and 4-5 show circuits that will be used to verify the following multivariable Boolean theorems.

X(Y + Z) = XY + XZ
X + XY = X
X + X'Y = X + Y

Figures 4-6 and 4-7 show circuits that will be used to verify the following DeMorgan's theorems.

(XY)' = X' + Y'
(X + Y)' = X'Y'

Figures 4-6 and 4-7 also show that the NAND gate is equivalent to an OR gate with INVERTERS on the inputs, and the NOR gate is equivalent to an AND gate with INVERTERS on the inputs.

Figure 4-1 Boolean Theorems—AND Operation

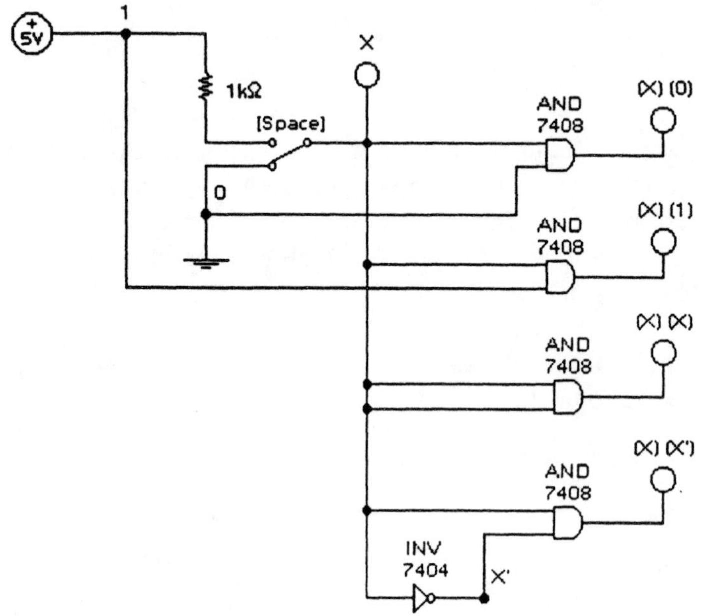

Figure 4-2 Boolean Theorems—OR Operation

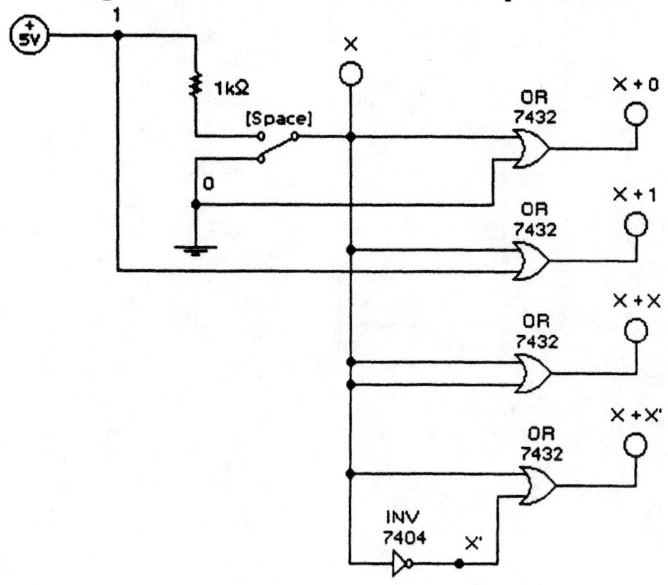

Figure 4-3 Multivariable Boolean Theorem

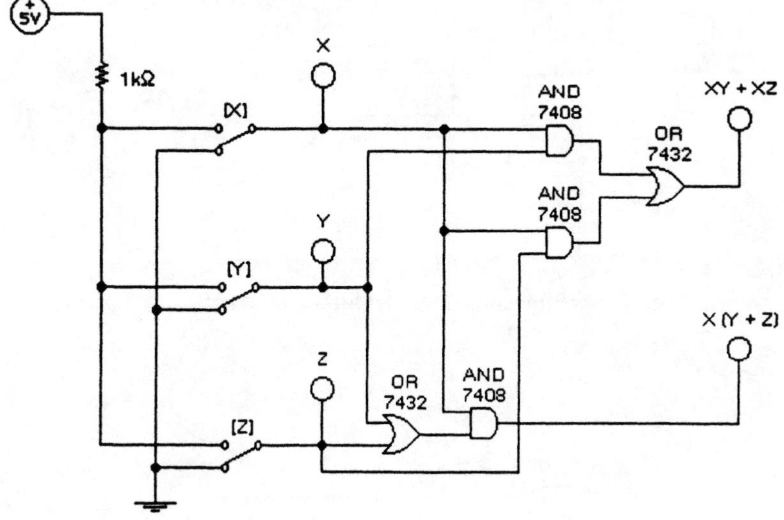

Figure 4-4 Multivariable Boolean Theorem

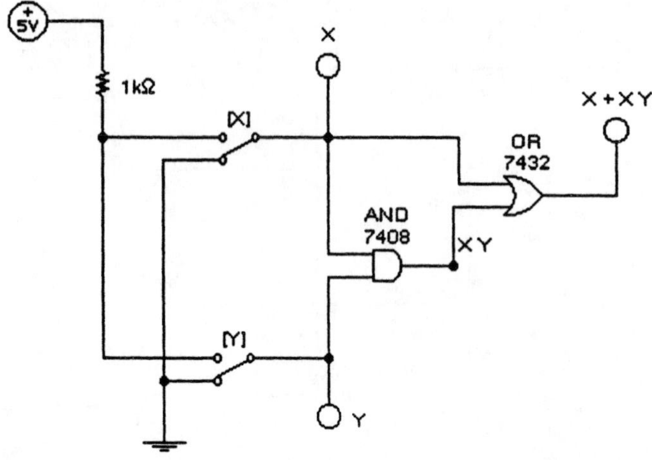

Figure 4-5 Multivariable Boolean Theorem

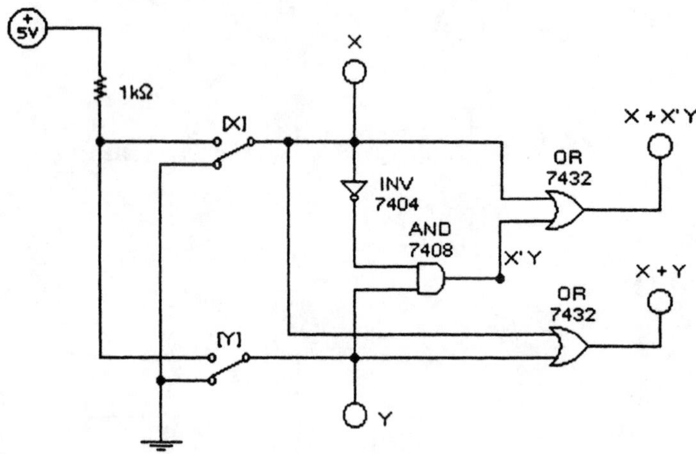

Figure 4-6 DeMorgan's Theorem

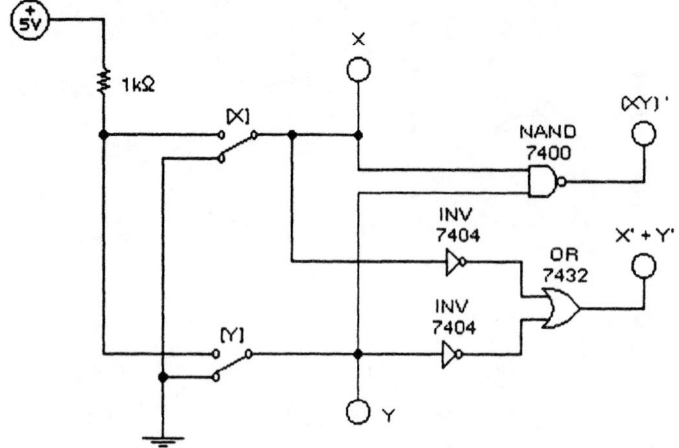

Figure 4-7 DeMorgan's Theorem

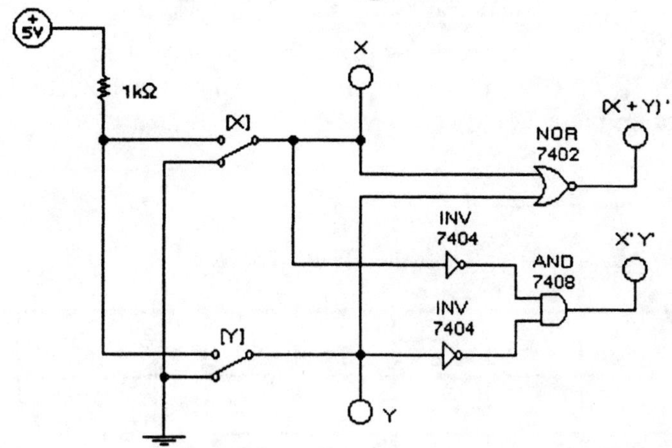

Procedure:

1. Pull down the File menu and open FIG4-1. Click the On-Off switch to run the analysis. Press the space bar on the keyboard to switch input X between a low (0) and a high (1) binary input to the AND gates. Observe the outputs as X is varied between binary zero (0) and binary one (1). Based on these results, complete the following Boolean identities.

   (X)(0) = ________ (X)(1) = ________

   (X)(X) = ________ (X)(X') = ________

Question: Do these results correspond with the known Boolean theorems?

2. Click the On-Off switch to stop the analysis. Pull down the File menu and open FIG4-2. Click the On-Off switch to run the analysis. Press the space bar on the keyboard to switch input X between a low (0) and a high (1) binary input to the OR gates. Observe the outputs as X is varied between binary zero (0) and binary one (1). Based on these results, complete the following Boolean identities.

   X + 0 = ________ X + 1 = ________

   X + X = ________ X + X' = ________

Question: Do these results correspond with the known Boolean theorems?

3. Click the On-Off switch to stop the analysis. Pull down the File menu and open FIG4-3. Click the On-Off switch to run the analysis. Press the "X", "Y" and "Z" keys on the keyboard to switch inputs X, Y, and Z between a low (0) and a high (1) binary input to the logic network. Record the outputs X(Y + Z) and XY + XZ for each input combination on the truth table (Table 4-1). Based on these results, complete the Boolean identity.

X(Y + Z) = ________

Table 4-1 Truth Table

X	Y	Z	X(Y + Z)	XY + XZ
0	0	0		
0	0	1		
0	1	0		
0	1	1		
1	0	0		
1	0	1		
1	1	0		
1	1	1		

Question: Do these results correspond with the known Boolean theorem?

4. Click the On-Off switch to stop the analysis. Pull down the File menu and open FIG4-4. Click the On-Off switch to run the analysis. Press the "X" and "Y" keys on the keyboard to switch inputs X and Y between a low (0) and a high (1) binary input to the logic network. Record the output X + XY for each input combination on the truth table (Table 4-2). Based on these results, complete the Boolean identity.

X + XY = ________

Table 4-2 Truth Table

X	Y	X + XY
0	0	
0	1	
1	0	
1	1	

Question: Do these results correspond with the known Boolean theorem?

5. Click the On-Off switch to stop the analysis. Pull down the File menu and open FIG4-5. Click the On-Off switch to run the analysis. Press the "X" and "Y" keys on the keyboard to switch inputs X and Y between a low (0) and a high (1) binary input to the logic network. Record the outputs X + X'Y and X + Y for each input combination on the truth table (Table 4-3). Based on these results, complete the Boolean identity.

   X + X'Y = ________

Table 4-3 Truth Table

X	Y	X + X'Y	X + Y
0	0		
0	1		
1	0		
1	1		

Question: Do these results correspond with the known Boolean theorem?

6. Click the On-Off switch to stop the analysis. Pull down the File menu and open FIG4-6. Click the On-Off switch to run the analysis. Press the "X" and "Y" keys on the keyboard to switch inputs X and Y between a low (0) and a high (1) binary input to the logic network. Record the outputs (XY)' and X' + Y' for each input combination on the truth table (Table 4-4). Based on these results, complete the Boolean identity.

   (XY)' = ________

Table 4-4 Truth Table

X	Y	(XY)'	X' + Y'
0	0		
0	1		
1	0		
1	1		

Question: Do these results correspond with DeMorgan's theorem?

7. Click the On-Off switch to stop the analysis. Pull down the File menu and open FIG4-7. Click the On-Off switch to run the analysis. Press the "X" and "Y" keys on the keyboard to switch inputs X and Y between a low (0) and a high (1) binary input to the logic network. Record the outputs (X + Y)' and X'Y' for each input combination on the truth table (Table 4-5). Based on these results, complete the Boolean identity.

   (X + Y)' = ________

Table 4-5 Truth Table

X	Y	(X + Y)'	X'Y'
0	0		
0	1		
1	0		
1	1		

Question: Do these results correspond with DeMorgan's theorem?

Name ______________________
Date ______________________

Experiment 5 Universality of NAND and NOR Gates

Objectives:

1. Show how an INVERTER can be implemented using a NAND gate.
2. Show how an AND gate can be implemented using NAND gates.
3. Show how an OR gate can be implemented using NAND gates.
4. Show how a NOR gate can be implemented using NAND gates.
5. Show how an INVERTER can be implemented using a NOR gate.
6. Show how an OR gate can be implemented using NOR gates.
7. Show how an AND gate can be implemented using NOR gates.
8. Show how a NAND gate can be implemented using NOR gates.

Materials:

One 5 V dc voltage supply
Two logic switches
Three logic probe lights
Four two-input NAND gates (1-7400 IC)
Four two-input NOR gates (1-7402 IC)
Two 1 kΩ resistors

Preparation:

The operation of a logic circuit consisting of a combination of logic gates can be described with a Boolean expression. The Boolean expression for the output of a logic circuit is obtained by first writing the Boolean expression for the output of each logic gate, working from the logic circuit input towards the logic circuit output. In the resulting Boolean expression, all AND operations are assumed to be performed before the OR operations unless an OR operation is surrounded by parentheses, in which case the OR operation is performed first.

See the Preparation section of Experiment 2 for information on INVERTERS, OR gates, and AND gates. See the Preparation section of Experiment 3 for information on truth tables, NAND gates, and NOR gates. See the Preparation section of Experiment 4 for information on the Boolean theorems and DeMorgan's theorems.

The circuits in Figures 5-1 through 5-4 show how to implement the INVERTER, AND gate, OR gate, and NOR gate using NAND gates. The circuits in Figures 5-5 through 5-8 show how to implement the INVERTER, OR gate, AND gate, and NAND gate using NOR gates.

Figure 5-1 NAND Gate INVERTER

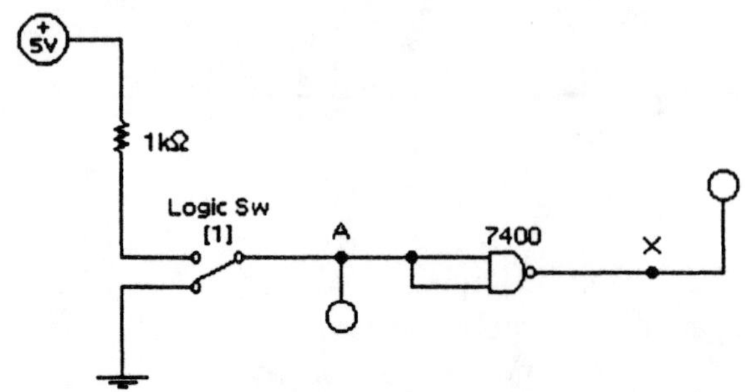

Figure 5-2 NAND Gate Implementation of an AND Gate

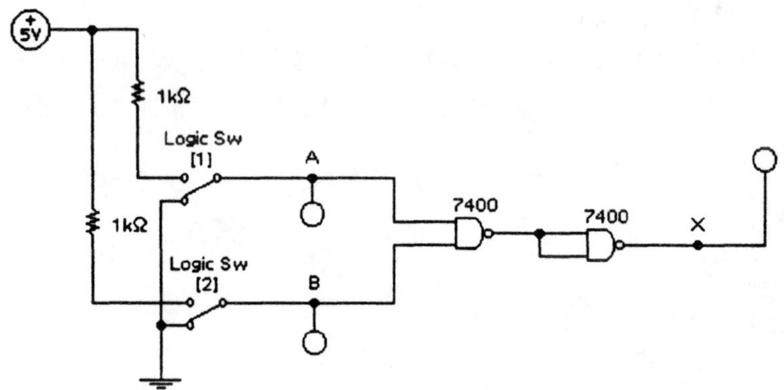

Figure 5-3 NAND Gate Implementation of an OR Gate

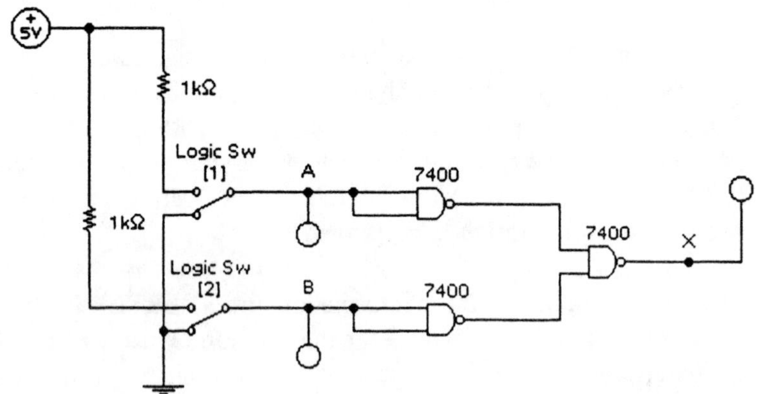

Figure 5-4 NAND Gate Implementation of a NOR Gate

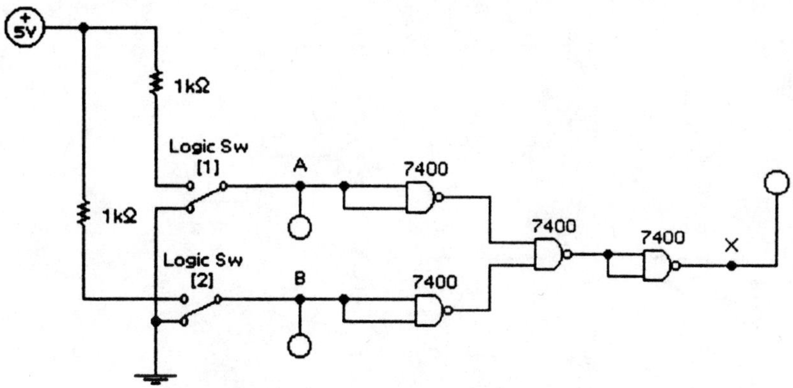

Figure 5-5 NOR Gate INVERTER

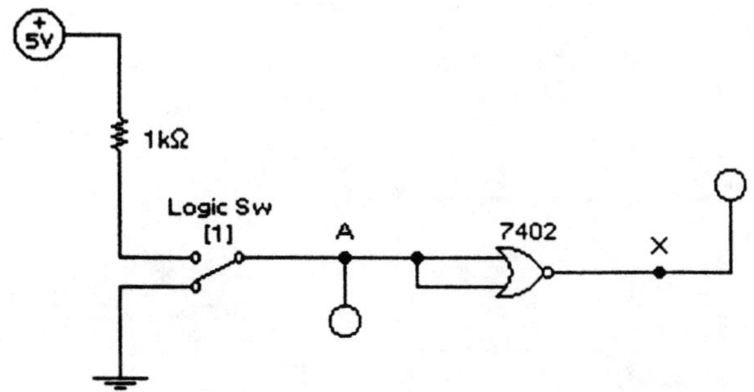

Figure 5-6 NOR Gate Implementation of an OR Gate

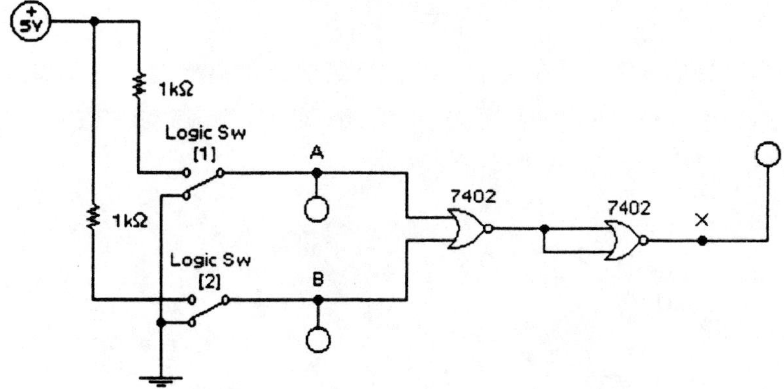

Figure 5-7 NOR Gate Implementation of an AND Gate

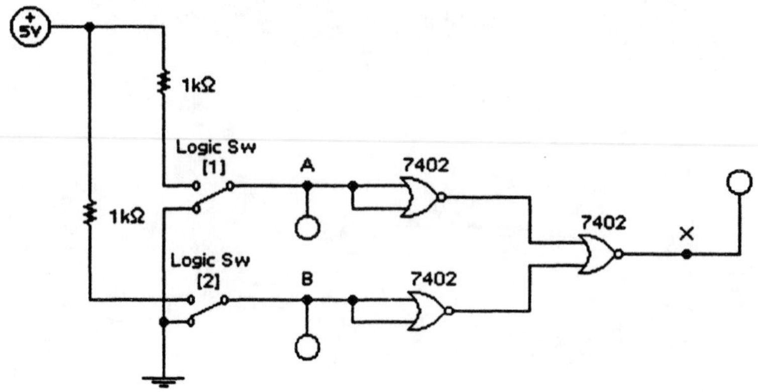

Figure 5-8 NOR Gate Implementation of a NAND Gate

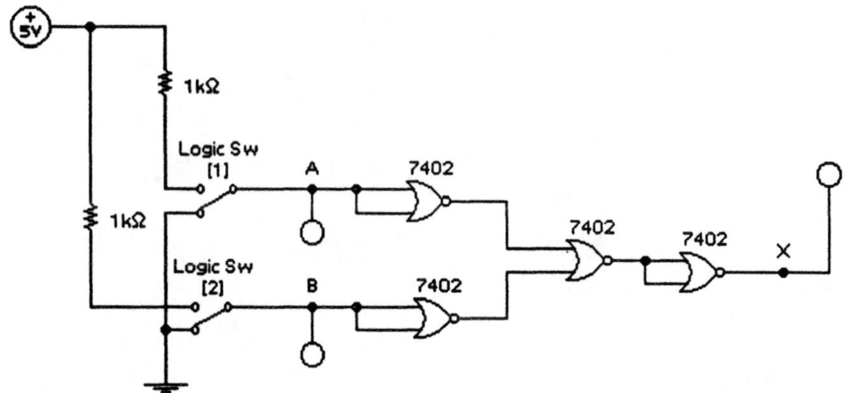

Procedure:

1. Pull down the File menu and open FIG5-1. Determine the Boolean expression for the logic circuit in Figure 5-1.

2. Click the On-Off switch to run the analysis. By switching the logic switch to the appropriate positions, complete the truth table (Table 5-1). The logic switch can be switched by pressing the "1" key on the keyboard.

Table 5-1 Truth Table

A	X
0	
1	

Question: Based on the logic equation in Step 1 and the truth table in Step 2, what is the logic circuit in Figure 5-1 equivalent to?

3. Click the On-Off switch to stop the analysis run. Pull down the File menu and open FIG5-2. Determine the Boolean expression for the logic circuit in Figure 5-2.

4. Click the On-Off switch to run the analysis. By switching the logic switches to the appropriate positions, complete the truth table (Table 5-2). The logic switches can be switched by pressing the "1" and "2" keys on the keyboard.

Table 5-2 Truth Table

A	B	X
0	0	
0	1	
1	0	
1	1	

Question: Based on the logic equation in Step 3 and the truth table in Step 4, what is the logic circuit in Figure 5-2 equivalent to?

5. Click the On-Off switch to stop the analysis run. Pull down the File menu and open FIG5-3. Determine the Boolean expression for the logic circuit in Figure 5-3.

6. Click the On-Off switch to run the analysis. By switching the logic switches to the appropriate positions, complete the truth table (Table 5-3). The logic switches can be switched by pressing the "1" and "2" keys on the keyboard.

Table 5-3 Truth Table

A	B	X
0	0	
0	1	
1	0	
1	1	

Question: Based on the logic equation in Step 5 and the truth table in Step 6, what is the logic circuit in Figure 5-3 equivalent to?

7. Click the On-Off switch to stop the analysis run. Pull down the File menu and open FIG5-4. Determine the Boolean expression for the logic circuit in Figure 5-4.

8. Click the On-Off switch to run the analysis. By switching the logic switches to the appropriate positions, complete the truth table (Table 5-4). The logic switches can be switched by pressing the "1" and "2" keys on the keyboard.

Table 5-4 Truth Table

A	B	X
0	0	
0	1	
1	0	
1	1	

Question: Based on the logic equation in Step 7 and the truth table in Step 8, what is the logic circuit in Figure 5-4 equivalent to?

9. Click the On-Off switch to stop the analysis run. Pull down the File menu and open FIG5-5. Determine the Boolean expression for the logic circuit in Figure 5-5.

10. Click the On-Off switch to run the analysis. By switching the logic switch to the appropriate positions, complete the truth (Table 5-5). The logic switch can be switched by pressing the "1" key on the keyboard.

Table 5-5 Truth Table

A	X
0	
1	

Question: Based on the logic equation in Step 9 and the truth table in Step 10, what is the logic circuit in Figure 5-5 equivalent to?

11. Click the On-Off switch to stop the analysis run. Pull down the File menu and open FIG5-6. Determine the Boolean expression for the logic circuit in Figure 5-6.

12. Click the On-Off switch to run the analysis. By switching the logic switches to the appropriate positions, complete the truth table (Table 5-6). The logic switches can be switched by pressing the "1" and "2" keys on the keyboard.

Table 5-6 Truth Table

A	B	X
0	0	
0	1	
1	0	
1	1	

Question: Based on the logic equation in Step 11 and the truth table in Step 12, what is the logic circuit in Figure 5-6 equivalent to?

13. Click the On-Off switch to stop the analysis run. Pull down the File menu and open FIG5-7. Determine the Boolean expression for the logic circuit in Figure 5-7.

14. Click the On-Off switch to run the analysis. By switching the logic switches to the appropriate positions, complete the truth table (Table 5-7). The logic switches can be switched by pressing the "1" and "2" keys on the keyboard.

Table 5-7 Truth Table

A	B	X
0	0	
0	1	
1	0	
1	1	

Question: Based on the logic equation in Step 13 and the truth table in Step 14, what is the logic circuit in Figure 5-7 equivalent to?

15. Click the On-Off switch to stop the analysis run. Pull down the File menu and open FIG5-8. Determine the Boolean expression for the logic circuit in Figure 5-8.

16. Click the On-Off switch to run the analysis. By switching the logic switches to the appropriate positions, complete the truth table (Table 5-8). The logic switches can be switched by pressing the "1" and "2" keys on the keyboard.

Table 5-8 Truth Table

A	B	X
0	0	
0	1	
1	0	
1	1	

Question: Based on the logic equation in Step 15 and the truth table in Step 16, what is the logic circuit in Figure 5-8 equivalent to?

Name ___________________________
Date ___________________________

Analyzing Combinational Logic Circuits

Objectives:

1. Investigate how to determine the Boolean expression for a logic circuit.
2. Investigate how to derive the truth table for a logic circuit.
3. Investigate how to convert AND-OR logic to NAND-NAND logic.
4. Investigate how to convert OR-AND logic to AND-OR logic.
5. Investigate how to convert NAND-NAND logic to AND-OR logic.

Materials:

One logic probe light
Three logic switches
Two two-input AND gates (1-7408 IC)
One two-input OR gate (1-7432 IC)
Three two-input NAND gates (1-7400 IC)
One INVERTER (1-7404 IC)

Preparation:

The operation of a logic circuit consisting of a combination of logic gates can be described with a Boolean expression. The Boolean expression for the output of a logic circuit is obtained by first writing the Boolean expression for the output of each logic gate, working from the logic circuit input towards the logic circuit output. In the resulting Boolean expression, all AND operations are assumed to be performed before the OR operations unless an OR operation is surrounded by parentheses, in which case the OR operation is performed first.

A truth table for a logic circuit shows the outputs for all possible input combinations. The truth table can be developed from the Boolean expression by evaluating the expression for all possible input combinations. This is accomplished by first evaluating each part of the expression for all possible input combinations, taking operations surrounded by parentheses first, AND operations next, and OR operations last. If an expression is inverted, perform the operation of the expression first, and then invert the result.

An AND-OR logic circuit can be easily converted to NAND-NAND logic by adding INVERTERS to the outputs of all the AND gates and all the inputs of the OR gate. This results in double inversion between each AND gate output and the OR gate inputs, resulting in no change to the circuit operation because double inversion is the same as no inversion. Because the OR gate with inversion at the inputs is equivalent to a NAND gate (DeMorgan's theorem), the resulting circuit consists of all NAND gates.

A NAND-NAND logic circuit can be converted to AND-OR logic by representing the last NAND gate as an OR gate with inversion on the inputs (DeMorgan's theorem). This will result in double inversion between the outputs of the NAND gates and the OR gate inputs. Remove the double INVERTERS and the AND-OR logic circuit will result.

Review the discussion of Figure 4-3 in the Preparation section of Experiment 4 to determine how to convert a product-of-sum (OR-AND) Boolean expression to a sum-of-products (AND-OR) Boolean expression. Review the discussion of Figures 4-6 and 4-7 and DeMorgan's theorem in the Preparation section of Experiment 4 to determine how to simplify a logic expression for a NAND-NAND logic circuit.

The circuits in Figures 6-1, 6-2, 6-3, and 6-4 will help demonstrate how to determine the Boolean expression for a logic circuit, how to derive a truth table, how to convert an AND-OR logic circuit to NAND-NAND logic, how to convert an OR-AND logic circuit to AND-OR logic, and how to convert a NAND-NAND logic circuit to AND-OR logic.

Figure 6-1 AND-OR Combinational Logic Network

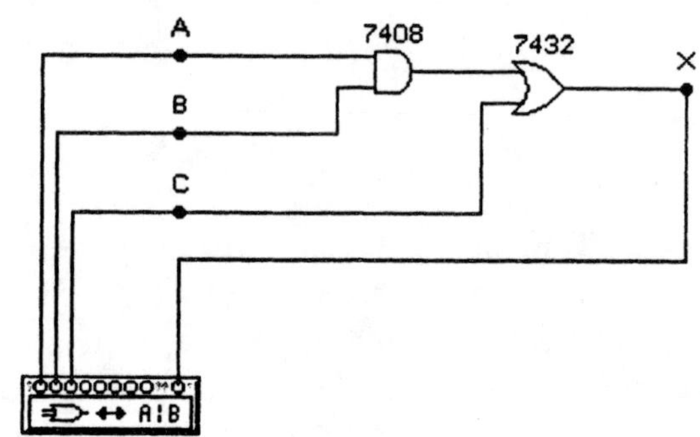

Figure 6-2 OR-AND Combinational Logic Network

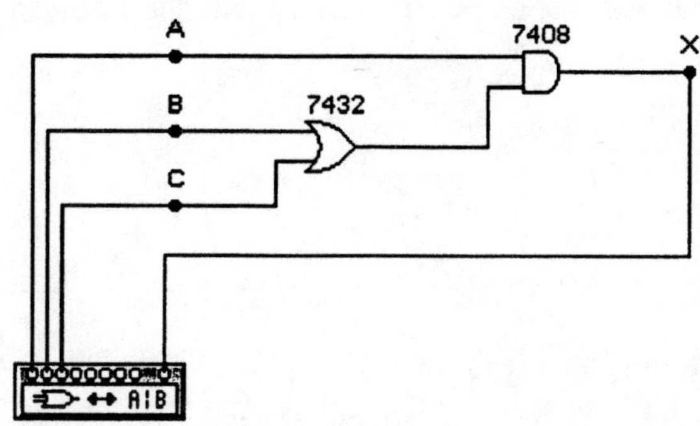

Figure 6-3 NAND-NAND Combinational Logic Network

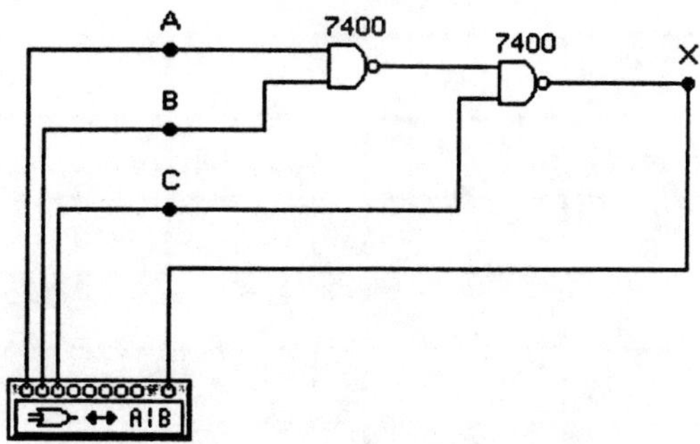

Figure 6-4 AND-OR Combinational Logic Network

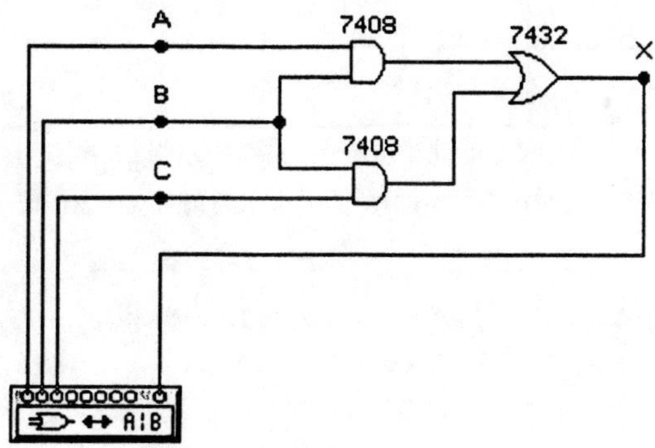

Procedure:

1. Pull down the File menu and open FIG6-1. Determine the Boolean expression for the logic circuit in Figure 6-1.

2. Complete the truth table (Table 6-1) for the logic circuit in Figure 6-1 from the Boolean expression in Step 1. Use the space to the right of the truth table to evaluate each part of the Boolean expression first.

Table 6-1 Truth Table

A	B	C	X		
0	0	0			
0	0	1			
0	1	0			
0	1	1			
1	0	0			
1	0	1			
1	1	0			
1	1	1			

3. Click the "Circuit to Truth Table" button on the logic converter.

> NOTE: A logic converter is not available in a hardwired laboratory. If you are performing this experiment in a hardwired laboratory environment, use logic switches for inputs A, B, and C and a logic probe light to monitor the output (X) to verify the truth table, in place of the logic converter.

Question: How does your truth table in Step 2 compare with the truth table on the logic converter (or verified on the hardwired circuit)?

4. Click the "Truth Table to Simplified Boolean Expression" button on the logic converter.

> NOTE: This step can't be performed in a hardwired laboratory environment.

Question: How does your Boolean expression in Step 1 compare with the Boolean expression on the logic converter?

5. Convert the AND-OR logic circuit shown in Figure 6-1 to NAND-NAND logic and draw the NAND-NAND equivalent circuit in the space provided.

6. Click the "Boolean Expression to NAND" button on the logic converter. Double click the upper left corner of the enlarged logic converter to remove the enlargement. Move the NAND-NAND logic circuit to the upper right hand corner of the workspace.

> NOTE: If you are performing this experiment in a hardwired laboratory, build the NAND-NAND logic circuit drawn in Step 5 and compare its truth table with the truth table for the original logic circuit (Table 6-1) in Step 2.

Question: How does the NAND-NAND logic circuit drawn in Step 5 compare with the NAND-NAND logic circuit shown in red on the screen (or how does the NAND-NAND truth table compare with the original truth table in Step 2)?

7. Pull down the File menu and open FIG6-2. Determine the Boolean expression for the logic circuit in Figure 6-2.

Question: How does this Boolean expression compare with the Boolean expression for the circuit in Figure 6-1 derived in Step 1? Explain.

8. Convert the Boolean expression in Step 7 to sum-of-products (AND-OR) form.

9. Complete the truth table (Table 6-2) for the logic circuit in Figure 6-2 from the Boolean expression in Step 8. Use the space to the right of the truth table to evaluate each part of the Boolean expression first.

Table 6-2 Truth Table

A	B	C	X		
0	0	0			
0	0	1			
0	1	0			
0	1	1			
1	0	0			
1	0	1			
1	1	0			
1	1	1			

10. Click the "Circuit to Truth Table" button on the logic converter.

NOTE: A logic converter is not available in a hardwired laboratory. If you are performing this experiment in a hardwired laboratory environment, use logic switches for inputs A, B, and C and a logic probe light to monitor the output (X) to verify the truth table, in place of the logic converter.

Question: How does your truth table in Step 9 compare with the truth table on the logic converter (or verified on the hardwired circuit)?

11. Click the "Truth Table to Simplified Boolean Expression" button on the logic converter.

NOTE: This step can't be performed in a hardwired laboratory environment.

Question: How does your AND-OR Boolean expression in Step 8 compare with the AND-OR Boolean expression on the logic converter?

12. Convert the logic circuit in Figure 6-2 to an AND-OR logic circuit from the sum-of-products expression derived in Step 8 and draw the AND-OR logic equivalent circuit in the space provided.

13. Click the "Boolean Expression to Circuit" button on the logic converter. Double click the upper left corner of the enlarged logic converter to remove the enlargement. Move the AND-OR logic circuit to the upper right hand corner of the workspace.

> NOTE: If you are performing this experiment in a hardwired laboratory, build the AND-OR logic circuit drawn in Step 12 and compare its truth table with the truth table for the original logic circuit (Table 6-2) in Step 9.

Question: How does the AND-OR logic circuit drawn in Step 12 compare with the AND-OR logic circuit shown in red on the screen (or how does the AND-OR truth table compare with the original truth table in Step 9)?

14. Convert the AND-OR logic circuit in Step 12 to NAND-NAND logic and draw the circuit in the space provided.

15. Double click the logic converter to bring down the enlargement again. Click the "Boolean Expression to NAND" button on the logic converter. Double click the upper left corner of the enlarged logic converter to remove the enlargement.

> **NOTE: If you are performing this experiment in a hardwired laboratory, build the NAND-NAND logic circuit drawn in Step 14 and compare its truth table with the truth table for the original logic circuit in Step 9.**

Question: How does the NAND-NAND logic circuit drawn in Step 14 compare with the logic circuit shown in red on the screen (or how does the NAND-NAND truth table compare with the original truth table in Step 9)?

16. Pull down the File menu and open FIG6-3. Determine the Boolean expression for the logic circuit in Figure 6-3.

17. Simplify the logic expression in Step 16 using DeMorgan's theorem.

18. Complete the truth table (Table 6-3) for the logic circuit in Figure 6-3 using the Boolean expression derived in Step 17. Use the space to the right of the truth table to evaluate each part of the Boolean expression first.

Table 6-3 Truth Table

A	B	C	X		
0	0	0			
0	0	1			
0	1	0			
0	1	1			
1	0	0			
1	0	1			
1	1	0			
1	1	1			

19. Click the "Circuit to Truth Table" button on the logic converter.

NOTE: A logic converter is not available in a hardwired laboratory. If you are performing this experiment in a hardwired laboratory environment, use logic switches for inputs A, B, and C and a logic probe light to monitor the output (X) to verify the truth table, in place of the logic converter.

Question: How does your truth table in Step 18 compare with the truth table on the logic converter (or verified on the hardwired circuit)?

20. Click the "Truth Table to Simplified Boolean Expression" button on the logic converter.

NOTE: This step can't be performed in a hardwired laboratory environment.

Question: How does your AND-OR Boolean expression in Step 17 compare with the AND-OR Boolean expression on the logic converter?

21. Convert the NAND-NAND logic circuit shown in Figure 6-3 to AND-OR logic and draw the equivalent circuit in the space provided.

22. Click the "Boolean Expression to Circuit" button on the logic converter. Double click the upper left corner of the enlarged logic converter to remove the enlargement. Move the AND-OR logic circuit to the upper right hand corner of the workspace.

NOTE: If you are performing this experiment in a hardwired laboratory, build the AND-OR logic circuit drawn in Step 21 and compare its truth table with the truth table for the original logic circuit (Table 6-3) in Step 18.

Question: How does the AND-OR logic circuit drawn in Step 21 compare with the logic circuit shown in red on the screen (or how does the AND-OR truth table compare with the original truth table in Step 18)?

23. Pull down the File menu and open FIG6-4. Determine the Boolean expression for the logic circuit in Figure 6-4.

24. Complete the truth table (Table 6-4) for the logic circuit in Figure 6-4 using the Boolean expression in Step 23. Use the space to the right of the truth table to evaluate each part of the Boolean expression first.

Table 6-4 Truth Table

A	B	C	X		
0	0	0			
0	0	1			
0	1	0			
0	1	1			
1	0	0			
1	0	1			
1	1	0			
1	1	1			

25. Click the "Circuit to Truth Table" button on the logic converter.

NOTE: A logic converter is not available in a hardwired laboratory. If you are performing this experiment in a hardwired laboratory environment, use logic switches for inputs A, B, and C and a logic probe light to monitor the output (X) to verify the truth table, in place of the logic converter.

Question: How does your truth table in Step 24 compare with the truth table on the logic converter (or verified on the hardwired circuit)?

26. Click the "Truth Table to Simplified Boolean Expression" button on the logic converter.

NOTE: This step can't be performed in a hardwired laboratory environment.

Question: How does your AND-OR Boolean expression in Step 23 compare with the AND-OR Boolean expression on the logic converter?

27. Convert the AND-OR logic circuit in Figure 6-4 to NAND-NAND logic and draw the circuit in the space provided.

28. Click the "Boolean Expression to NAND" button on the logic converter. Double click the upper left corner of the enlarged logic converter to remove the enlargement. Move the NAND-NAND logic circuit to the upper right hand corner of the workspace.

NOTE: If you are performing this experiment in a hardwired laboratory, build the NAND-NAND logic circuit drawn in Step 27 and compare its truth table with the truth table for the original logic circuit (Table 6-4) in Step 24.

Question: How does the NAND-NAND logic circuit drawn in Step 27 compare with the logic circuit shown in red on the screen (or how does the NAND-NAND truth table compare with the original truth table in Step 24)?

Name ______________________________

Date ______________________________

Simplifying Combinational Logic Circuits

Objectives:

1. Simplify combinational logic circuits using the Boolean theorems.
2. Simplify combinational logic circuits using DeMorgan's theorems.

Materials:

One 5 V dc power supply
Three logic switches
One logic probe light
Two two-input AND gates (1-7408 IC)
Two two-input OR gates (1-7432 IC)
Two INVERTERS (1-7404 IC)
One three-input AND gate (1-7411 IC)
One two-input NAND gate (1-7400 IC)
One two-input NOR gate (1-7402 IC)

Preparation:

Review the Preparation section of Experiment 6 for writing the Boolean expression for a combinational logic circuit. Review the Preparation section of Experiment 4 for the Boolean theorems and Demorgan's theorems. In this experiment, you will write the Boolean expressions for the logic circuits in Figures 7-1, 7-2, 7-3, 7-4, 7-5, and 7-6. Then you will simplify the Boolean expressions using the Boolean theorems and DeMorgan's theorems and draw the simplified logic circuits from the simplified Boolean expressions.

Figure 7-1 AND-OR Combinational Logic Circuit

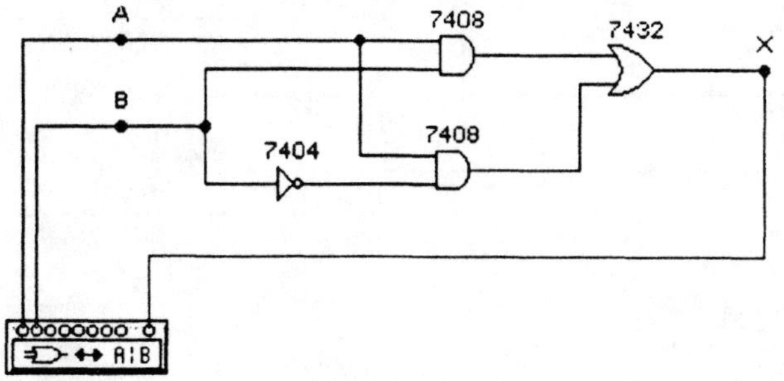

Figure 7-2 AND-OR Combinational Logic Circuit

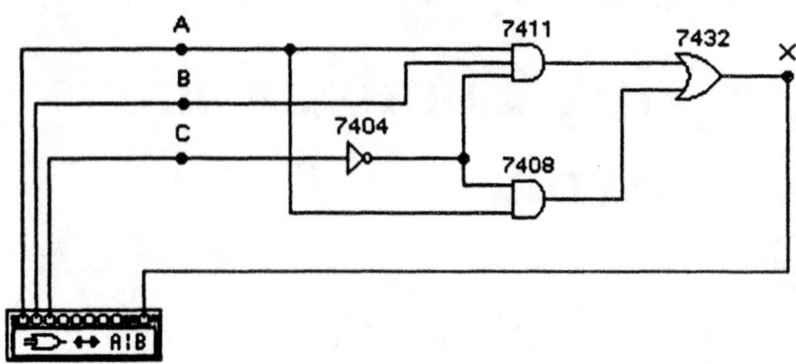

Figure 7-3 OR-AND Combinational Logic Circuit

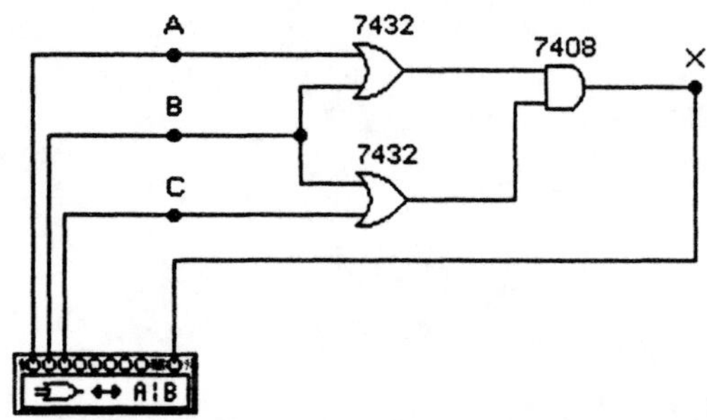

Figure 7-4 Combinational Logic Circuit

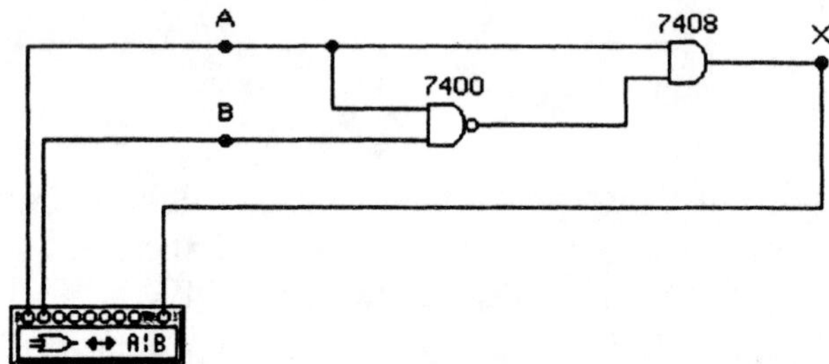

Figure 7-5 Combinational Logic Circuit

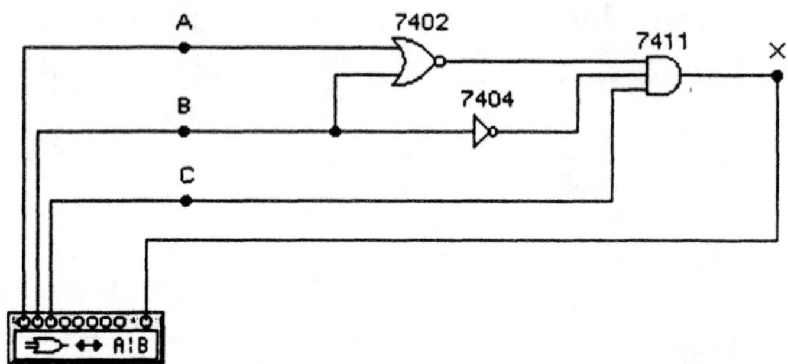

Figure 7-6 Combinational Logic Circuit

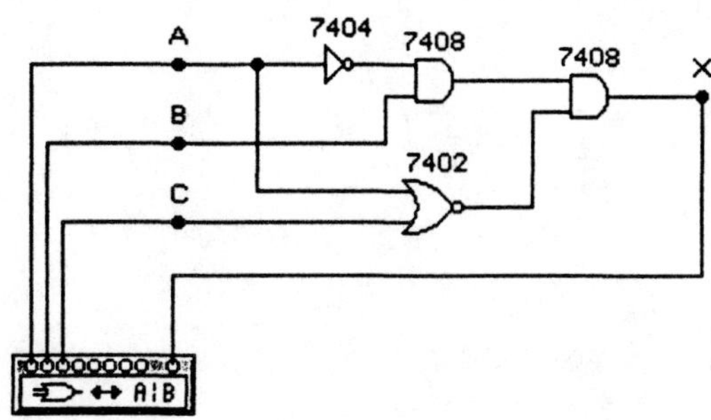

Procedure:

1. Pull down the File menu and open FIG7-1. Write the Boolean expression for the logic circuit in Figure 7-1.

2. Click the "Circuit to Truth Table" button on the logic converter.

> NOTE: A logic converter is not available in a hardwired laboratory. If you are performing this experiment in a hardwired laboratory, build the logic circuit in Figure 7-1, use logic switches for inputs A and B, and use a logic probe light to monitor output (X). Measure the outputs for each input combination to complete the truth table (Table 7-1) for the logic circuit.

Table 7-1 Truth Table

A	B	X
0	0	
0	1	
1	0	
1	1	

3. Simplify the Boolean expression in Step 1 using the Boolean theorems.

4. Click the "Truth Table to Simplified Boolean Expression" button on the logic converter.

> NOTE: This step can't be performed in a hardwired laboratory environment.

Question: How does your simplified Boolean expression in Step 3 compare with the Boolean expression on the logic converter?

5. Draw the simplified logic circuit from the simplified Boolean expression in Step 3.

6. Click the "Boolean Expression to Logic Circuit" button on the logic converter. Double click the logic converter in the upper left hand corner to reduce the enlargement.

> NOTE: A logic converter is not available in a hardwired laboratory. If you are performing this experiment in a hardwired laboratory, build the simplified logic circuit in Step 5, use logic switches for inputs A and B, and use a logic probe light to monitor output (X). Measure the outputs for each input combination to complete the truth table (Table 7-2) for the simplified logic circuit.

Table 7-2 Truth Table

A	B	X
0	0	
0	1	
1	0	
1	1	

Question: How does your simplified logic circuit in Step 5 compare with the logic circuit in red on the screen (or how does the truth table for the simplified logic circuit compare with the truth table for the original circuit)?

7. Pull down the File menu and open FIG7-2. Write the Boolean expression for the logic circuit in Figure 7-2.

8. Click the "Circuit to Truth Table" button on the logic converter.

> NOTE: A logic converter is not available in a hardwired laboratory. If you are performing this experiment in a hardwired laboratory, build the logic circuit in Figure 7-2, use logic switches for inputs A, B, and C, and use a logic probe light to monitor output (X). Measure the outputs for each input combination to complete the truth table (Table 7-3) for the logic circuit.

Table 7-3 Truth Table

A	B	C	X
0	0	0	
0	0	1	
0	1	0	
0	1	1	
1	0	0	
1	0	1	
1	1	0	
1	1	1	

9. Simplify the Boolean expression in Step 7 using the Boolean theorems.

10. Click the "Truth Table to Simplified Boolean Expression" button on the logic converter.

NOTE: This step can't be performed in a hardwired laboratory environment.

Question: How does your simplified Boolean expression in Step 9 compare with the Boolean expression on the logic converter?

11. Draw the simplified logic circuit from the simplified Boolean expression in Step 9.

12. Click the "Boolean Expression to Logic Circuit" button on the logic converter. Double click the logic converter in the upper left hand corner to reduce the enlargement.

NOTE: A logic converter is not available in a hardwired laboratory. If you are performing this experiment in a hardwired laboratory, build the simplified logic circuit in Step 11, use logic switches for inputs A, B, and C, and use a logic probe light to monitor output (X). Measure the outputs for each input combination to complete the truth table (Table 7-4) for the simplified logic circuit.

Table 7-4 Truth Table

A	B	C	X
0	0	0	
0	0	1	
0	1	0	
0	1	1	
1	0	0	
1	0	1	
1	1	0	
1	1	1	

Question: How does your simplified logic circuit in Step 11 compare with the logic circuit in red on the screen (or how does the truth table for the simplified logic circuit compare with the truth table for the original circuit)?

13. Pull down the File menu and open FIG7-3. Write the Boolean expression for the logic circuit in Figure 7-3.

14. Click the "Circuit to Truth Table" button on the logic converter.

NOTE: A logic converter is not available in a hardwired laboratory. If you are performing this experiment in a hardwired laboratory, build the logic circuit in Figure 7-3, use logic switches for inputs A, B, and C, and use a logic probe light to monitor output (X). Measure the outputs for each input combination to complete the truth table (Table 7-5) for the logic circuit.

Table 7-5 Truth Table

A	B	C	X
0	0	0	
0	0	1	
0	1	0	
0	1	1	
1	0	0	
1	0	1	
1	1	0	
1	1	1	

15. Simplify the Boolean expression in Step 13 using the Boolean theorems. Write your answer in sum-of-products (AND-OR) form.

16. Click the "Truth Table to Simplified Boolean Expression" button on the logic converter.

NOTE: This step can't be performed in a hardwired laboratory environment.

Question: How does your simplified Boolean expression in Step 15 compare with the Boolean expression on the logic converter?

17. Draw the simplified logic circuit from the simplified Boolean expression in Step 15.

18. Click the "Boolean Expression to Logic Circuit" button on the logic converter. Double click the logic converter in the upper left hand corner to reduce the enlargement.

> NOTE: A logic converter is not available in a hardwired laboratory. If you are performing this experiment in a hardwired laboratory, build the simplified logic circuit in Step 17, use logic switches for inputs A, B, and C, and use a logic probe light to monitor output (X). Measure the outputs for each input combination to complete the truth table (Table 7-6) for the simplified logic circuit.

Table 7-6 Truth Table

A	B	C	X
0	0	0	
0	0	1	
0	1	0	
0	1	1	
1	0	0	
1	0	1	
1	1	0	
1	1	1	

Question: How does your simplified logic circuit in Step 17 compare with the logic circuit in red on the screen (or how does the truth table for the simplified logic circuit compare with the truth table for the original circuit)?

19. Pull down the File menu and open FIG7-4. Write the Boolean expression for the logic circuit in Figure 7-4.

20. Click the "Circuit to Truth Table" button on the logic converter.

NOTE: A logic converter is not available in a hardwired laboratory. If you are performing this experiment in a hardwired laboratory, build the logic circuit in Figure 7-4, use logic switches for inputs A and B, and use a logic probe light to monitor output (X). Measure the outputs for each input combination to complete the truth table (Table 7-7) for the logic circuit.

Table 7-7 Truth Table

A	B	X
0	0	
0	1	
1	0	
1	1	

21. Simplify the Boolean expression in Step 19 using the Boolean theorems and DeMorgan's theorem.

22. Click the "Truth Table to Simplified Boolean Expression" button on the logic converter.

NOTE: This step can't be performed in a hardwired laboratory environment.

Question: How does your simplified Boolean expression in Step 21 compare with the Boolean expression on the logic converter?

23. Draw the simplified logic circuit from the simplified Boolean expression in Step 21.

24. Click the "Boolean Expression to Logic Circuit" button on the logic converter. Double click the logic converter in the upper left hand corner to reduce the enlargement.

NOTE: A logic converter is not available in a hardwired laboratory. If you are performing this experiment in a hardwired laboratory, build the simplified logic circuit in Step 23, use logic switches for inputs A and B, and use a logic probe light to monitor output (X). Measure the outputs for each input combination to complete the truth table (Table 7-8) for the simplified logic circuit.

Table 7-8 Truth Table

A	B	X
0	0	
0	1	
1	0	
1	1	

Question: How does your simplified logic circuit in Step 23 compare with the logic circuit in red on the screen (or how does the truth table for the simplified logic circuit compare with the truth table for the original circuit)?

25. Pull down the File menu and open FIG7-5. Write the Boolean expression for the logic circuit in Figure 7-5.

26. Click the "Circuit to Truth Table" button on the logic converter.

NOTE: A logic converter is not available in a hardwired laboratory. If you are performing this experiment in a hardwired laboratory, build the logic circuit in Figure 7-5, use logic switches for inputs A, B, and C, and use a logic probe light to monitor output (X). Measure the outputs for each input combination to complete the truth table (Table 7-9) for the logic circuit.

Table 7-9 Truth Table

A	B	C	X
0	0	0	
0	0	1	
0	1	0	
0	1	1	
1	0	0	
1	0	1	
1	1	0	
1	1	1	

27. Simplify the Boolean expression in Step 25 using the Boolean theorems and DeMorgan's theorems.

28. Click the "Truth Table to Simplified Boolean Expression" button on the logic converter.

NOTE: This step can't be performed in a hardwired laboratory environment.

Question: How does your simplified Boolean expression in Step 27 compare with the Boolean expression on the logic converter?

29. Draw the simplified logic circuit from the simplified Boolean expression in Step 27.

30. Click the "Boolean Expression to Logic Circuit" button on the logic converter. Double click the logic converter in the upper left hand corner to reduce the enlargement.

NOTE: A logic converter is not available in a hardwired laboratory. If you are performing this experiment in a hardwired laboratory, build the simplified logic circuit in Step 29, use logic switches for inputs A, B, and C, and use a logic probe light to monitor output (X). Measure the outputs for each input combination to complete the truth table (Table 7-10) for the simplified logic circuit.

Table 7-10 Truth Table

A	B	C	X
0	0	0	
0	0	1	
0	1	0	
0	1	1	
1	0	0	
1	0	1	
1	1	0	
1	1	1	

Question: How does your simplified logic circuit in Step 29 compare with the logic circuit in red on the screen (or how does the truth table for the simplified logic circuit compare with the truth table for the original circuit)?

31. Pull down the File menu and open FIG7-6. Write the Boolean expression for the logic circuit in Figure 7-6.

32. Click the "Circuit to Truth Table" button on the logic converter.

NOTE: A logic converter is not available in a hardwired laboratory. If you are performing this experiment in a hardwired laboratory, build the logic circuit in Figure 7-6, use logic switches for inputs A, B, and C, and use a logic probe light to monitor output (X). Measure the outputs for each input combination to complete the truth table (Table 7-11) for the logic circuit.

Table 7-11 Truth Table

A	B	C	X
0	0	0	
0	0	1	
0	1	0	
0	1	1	
1	0	0	
1	0	1	
1	1	0	
1	1	1	

33. Simplify the Boolean expression in Step 31 using the Boolean theorems and DeMorgan's theorems.

34. Click the "Truth Table to Simplified Boolean Expression" button on the logic converter.

NOTE: This step can't be performed in a hardwired laboratory environment.

Question: How does your simplified Boolean expression in Step 33 compare with the Boolean expression on the logic converter?

35. Draw the simplified logic circuit from the simplified Boolean expression in Step 33.

36. Click the "Boolean Expression to Logic Circuit" button on the logic converter. Double click the logic converter in the upper left hand corner to reduce the enlargement.

NOTE: A logic converter is not available in a hardwired laboratory. If you are performing this experiment in a hardwired laboratory, build the simplified logic circuit in Step 35, use logic switches for inputs A, B, and C, and use a logic probe light to monitor output (X). Measure the outputs for each input combination to complete the truth table (Table 7-12) for the simplified logic circuit.

Table 7-12 Truth Table

A	B	C	X
0	0	0	
0	0	1	
0	1	0	
0	1	1	
1	0	0	
1	0	1	
1	1	0	
1	1	1	

Question: How does your simplified logic circuit in Step 35 compare with the logic circuit in red on the screen (or how does the truth table for the simplified logic circuit compare with the truth table for the original circuit)?

Name ______________________________
Date ______________________________

Experiment 8 Logic Simplification Using Karnaugh Maps

Objectives:

1. Simplify AND-OR combinational logic circuits using K-maps.

Materials:

One 5 V dc power supply
Four logic switches
One logic probe light
Two two-input AND gates (1-7408 IC)
Two two-input OR gates (1-7432 IC)
Three INVERTERS (1-7404 IC)
Three three-input AND gates (1-7411 IC)
One four-input AND gate (1-7421 IC)

Preparation:

Review the Preparation section of Experiment 6 on writing Boolean expressions for combinational logic circuits. In this experiment, you will write the Boolean expressions for the logic circuits in Figures 8-1, 8-2, 8-3, 8-4, 8-5, and 8-6. Then you will simplify the Boolean expressions using Karnaugh maps and draw the simplified logic circuits from the simplified Boolean expressions. Karnaugh maps provide a cookbook method of simplifying logic circuits.

Figure 8-1 AND-OR Combinational Logic Circuit

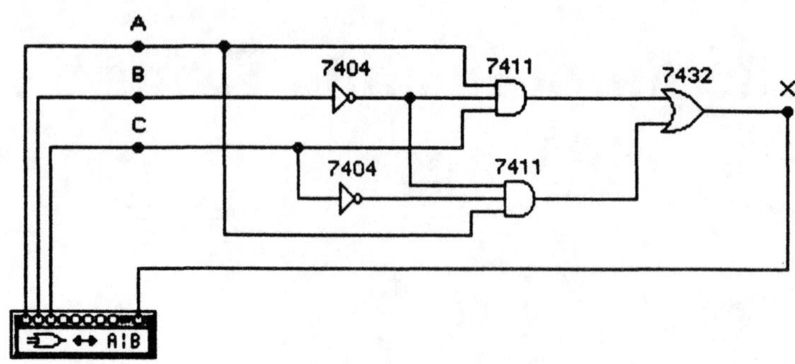

Figure 8-2 AND-OR Combinational Logic Circuit

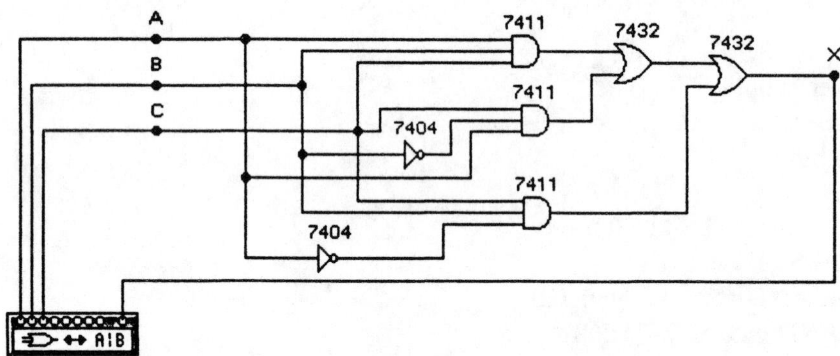

Figure 8-3 AND-OR Combinational Logic Circuit

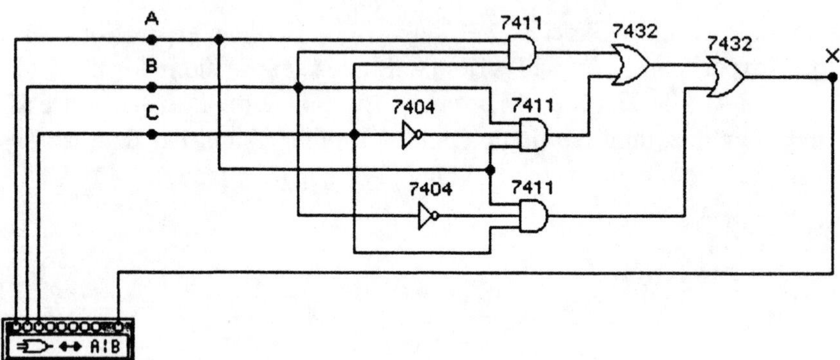

Figure 8-4 AND-OR Combinational Circuit

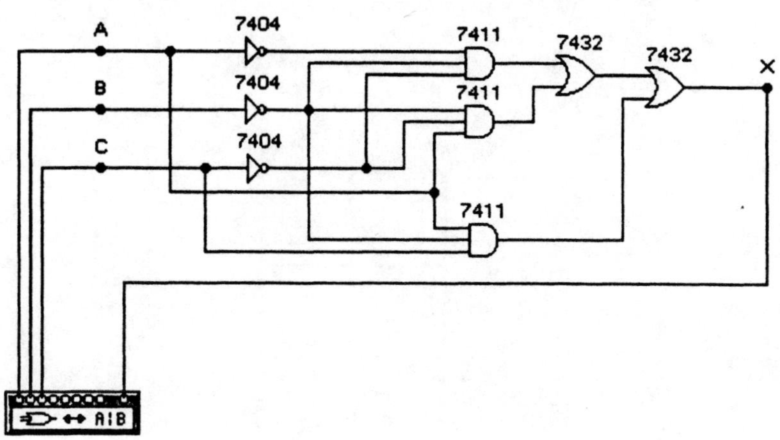

Figure 8-5 AND-OR Combinational Logic Circuit

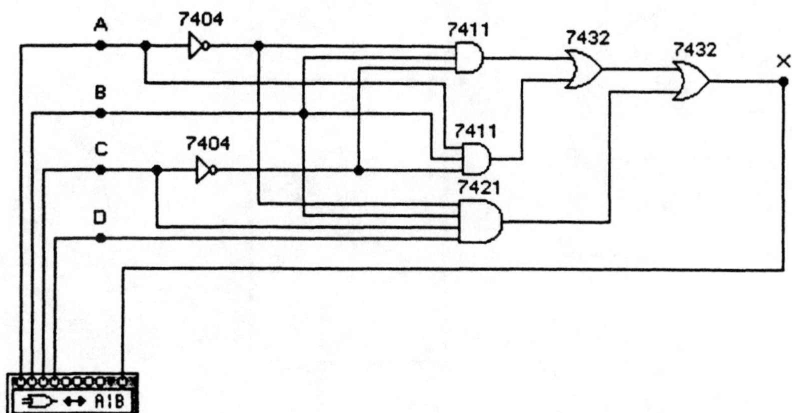

Figure 8-6 AND-OR Combinational Circuit

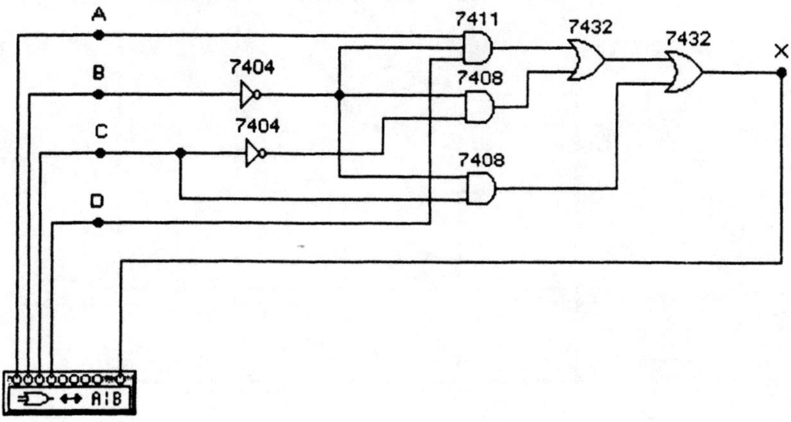

The 2-input K-map is shown in Figure 8-A, the 3-input K-map is shown in Figure 8-B, and the 4-input K-map is shown in Figure 8-C.

Figure 8-A Two-Input K-Map

	$\overline{B}$	B
$\overline{A}$		
A		

Figure 8-B Three-Input K-Map

	$\overline{C}$	C
$\overline{A}\,\overline{B}$		
$\overline{A}B$		
AB		
$A\overline{B}$		

Figure 8-C Four-Input K-Map

	$\overline{C}\,\overline{D}$	$\overline{C}D$	CD	$C\overline{D}$
$\overline{A}\,\overline{B}$				
$\overline{A}B$				
AB				
$A\overline{B}$				

When using a Karnaugh map to simplify a logic network, you must first write the network logic equation in sum-of-products (AND-OR) form. Then write a "1" in each K-map block that corresponds to each product term (AND operation) in your logic equation. In some cases the product term will take more than one block in the K-map. This will happen when the number of terms in the product term is less than the number of input variables.

After the K-map is filled with 1's in the appropriate blocks, encircle the adjacent 1's in groups of eight, four, and two, encircling the largest groups first. Consider the sides of the K-map to be adjacent to each other and the top and bottom of the K-map to be adjacent to each other. Overlapping of groups is allowed. Next, encircle any isolated 1's. Make sure that all of the 1's are encircled or within a grouping. Remove any circles that can be removed without exposing any 1's.

Each product term in the simplified sum-of-products expression consists of those variables that remain unchanged within each encircled group. Add any product terms representing isolated 1's. These product terms cannot be simplified.

Procedure:

1. Pull down the File menu and open FIG8-1. Write the Boolean expression for the logic circuit in Figure 8-1.

2. Click the "Circuit to Truth Table" button on the logic converter.

> NOTE: A logic converter is not available in a hardwired laboratory. If you are performing this experiment in a hardwired laboratory, build the logic circuit in Figure 8-1, use logic switches for inputs A, B, and C, and use a logic probe light to monitor output (X). Measure the outputs for each input combination to complete the truth table (Table 8-1) for the logic circuit.

Table 8-1 Truth Table

A	B	C	X
0	0	0	
0	0	1	
0	1	0	
0	1	1	
1	0	0	
1	0	1	
1	1	0	
1	1	1	

3. Draw the 3-input K-map and use it to simplify the Boolean expression in Step 1. Write the simplified Boolean expression.

4. Click the "Truth Table to Simplified Boolean Expression" button on the logic converter.

NOTE: This step can't be performed in a hardwired laboratory environment.

Question: How does your simplified Boolean expression in Step 3 compare with the Boolean expression on the logic converter?

5. Draw the simplified logic circuit from the simplified Boolean expression in Step 3.

6. Click the "Boolean Expression to Logic Circuit" button on the logic converter. Double click the logic converter in the upper left hand corner to reduce the enlargement.

NOTE: A logic converter is not available in a hardwired laboratory. If you are performing this experiment in a hardwired laboratory, build the simplified logic circuit in Step 5, use logic switches for inputs A, B, and C, and use a logic probe light to monitor output (X). Measure the outputs for each input combination to complete the truth table (Table 8-2) for the simplified logic circuit.

Table 8-2 Truth Table

A	B	C	X
0	0	0	
0	0	1	
0	1	0	
0	1	1	
1	0	0	
1	0	1	
1	1	0	
1	1	1	

Question: How does your simplified logic circuit in Step 5 compare with the logic circuit in red on the screen (or how does the truth table for the simplified logic circuit compare with the truth table for the original circuit)?

7. Pull down the File menu and open FIG8-2. Write the Boolean expression for the logic circuit in Figure 8-2.

8. Click the "Circuit to Truth Table" button on the logic converter.

NOTE: A logic converter is not available in a hardwired laboratory. If you are performing this experiment in a hardwired laboratory, build the logic circuit in Figure 8-2, use logic switches for inputs A, B, and C, and use a logic probe light to monitor output (X). Measure the outputs for each input combination to complete the truth table (Table 8-3) for the logic circuit.

Table 8-3 Truth Table

A	B	C	X
0	0	0	
0	0	1	
0	1	0	
0	1	1	
1	0	0	
1	0	1	
1	1	0	
1	1	1	

9. **Draw the 3-input K-map and use it to simplify the Boolean expression in Step 7. Write the simplified Boolean expression.**

10. Click the "Truth Table to Simplified Boolean Expression" button on the logic converter.

NOTE: This step can't be performed in a hardwired laboratory environment.

Question: How does your simplified Boolean expression in Step 9 compare with the Boolean expression on the logic converter?

11. Draw the simplified logic circuit from the simplified Boolean expression in Step 9.

12. Click the "Boolean Expression to Logic Circuit" button on the logic converter. Double click the logic converter in the upper left hand corner to reduce the enlargement.

NOTE: A logic converter is not available in a hardwired laboratory. If you are performing this experiment in a hardwired laboratory, build the simplified logic circuit in Step 11, use logic switches for inputs A, B, and C, and use a logic probe light to monitor output (X). Measure the outputs for each input combination to complete the truth table (Table 8-4) for the simplified logic circuit.

Table 8-4 Truth Table

A	B	C	X
0	0	0	
0	0	1	
0	1	0	
0	1	1	
1	0	0	
1	0	1	
1	1	0	
1	1	1	

Question: How does your simplified logic circuit in Step 11 compare with the logic circuit in red on the screen (or how does the truth table for the simplified logic circuit compare with the truth table for the original circuit)?

13. Pull down the File menu and open FIG8-3. Write the Boolean expression for the logic circuit in Figure 8-3.

14. Click the "Circuit to Truth Table" button on the logic converter.

NOTE: A logic converter is not available in a hardwired laboratory. If you are performing this experiment in a hardwired laboratory, build the logic circuit in Figure 8-3, use logic switches for inputs A, B, and C, and use a logic probe light to monitor output (X). Measure the outputs for each input combination to complete the truth table (Table 8-5) for the logic circuit.

Table 8-5 Truth Table

A	B	C	X
0	0	0	
0	0	1	
0	1	0	
0	1	1	
1	0	0	
1	0	1	
1	1	0	
1	1	1	

15. Draw the 3-input K-map and use it to simplify the Boolean expression in Step 13. Write the simplified Boolean expression.

16. Click the "Truth Table to Simplified Boolean Expression" button on the logic converter.

NOTE: This step can't be performed in a hardwired laboratory environment.

Question: How does your simplified Boolean expression in Step 15 compare with the Boolean expression on the logic converter?

17. Draw the simplified logic circuit from the simplified Boolean expression in Step 15.

18. Click the "Boolean Expression to Logic Circuit" button on the logic converter. Double click the logic converter in the upper left hand corner to reduce the enlargement.

NOTE: A logic converter is not available in a hardwired laboratory. If you are performing this experiment in a hardwired laboratory, build the simplified logic circuit in Step 17, use logic switches for inputs A, B, and C, and use a logic probe light to monitor output (X). Measure the outputs for each input combination to complete the truth table (Table 8-6) for the simplified logic circuit.

Table 8-6 Truth Table

A	B	C	X
0	0	0	
0	0	1	
0	1	0	
0	1	1	
1	0	0	
1	0	1	
1	1	0	
1	1	1	

Question: How does your simplified logic circuit in Step 17 compare with the logic circuit in red on the screen (or how does the truth table for the simplified logic circuit compare with the truth table for the original circuit)?

19. Pull down the File menu and open FIG8-4. Write the Boolean expression for the logic circuit in Figure 8-4.

20. Click the "Circuit to Truth Table" button on the logic converter.

NOTE: A logic converter is not available in a hardwired laboratory. If you are performing this experiment in a hardwired laboratory, build the logic circuit in Figure 8-4, use logic switches for inputs A, B, and C, and use a logic probe light to monitor output (X). Measure the outputs for each input combination to complete the truth table (Table 8-7) for the logic circuit.

Table 8-7 Truth Table

A	B	C	X
0	0	0	
0	0	1	
0	1	0	
0	1	1	
1	0	0	
1	0	1	
1	1	0	
1	1	1	

21. Draw the 3-input K-map and use it to simplify the Boolean expression in Step 19. Write the simplified Boolean expression.

22. Click the "Truth Table to Simplified Boolean Expression" button on the logic converter.

NOTE: This step can't be performed in a hardwired laboratory environment.

Question: How does your simplified Boolean expression in Step 21 compare with the Boolean expression on the logic converter?

23. Draw the simplified logic circuit from the simplified Boolean expression in Step 21.

24. Click the "Boolean Expression to Logic Circuit" button on the logic converter. Double click the logic converter in the upper left hand corner to reduce the enlargement.

NOTE: A logic converter is not available in a hardwired laboratory. If you are performing this experiment in a hardwired laboratory, build the simplified logic circuit in Step 23, use logic switches for inputs A, B, and C, and use a logic probe light to monitor output (X). Measure the outputs for each input combination to complete the truth table (Table 8-8) for the simplified logic circuit.

Table 8-8 Truth Table

A	B	C	X
0	0	0	
0	0	1	
0	1	0	
0	1	1	
1	0	0	
1	0	1	
1	1	0	
1	1	1	

Question: How does your simplified logic circuit in Step 23 compare with the logic circuit in red on the screen (or how does the truth table for the simplified logic circuit compare with the truth table for the original circuit)?

25. Pull down the File menu and open FIG8-5. Write the Boolean expression for the logic circuit in Figure 8-5.

26. Click the "Circuit to Truth Table" button on the logic converter.

NOTE: A logic converter is not available in a hardwired laboratory. If you are performing this experiment in a hardwired laboratory, build the logic circuit in Figure 8-5, use logic switches for inputs A, B, C, and D, and use a logic probe light to monitor output (X). Measure the outputs for each input combination to complete the truth table (Table 8-9) for the logic circuit.

Table 8-9 Truth Table

A	B	C	D	X
0	0	0	0	
0	0	0	1	
0	0	1	0	
0	0	1	1	
0	1	0	0	
0	1	0	1	
0	1	1	0	
0	1	1	1	
1	0	0	0	
1	0	0	1	
1	0	1	0	
1	0	1	1	
1	1	0	0	
1	1	0	1	
1	1	1	0	
1	1	1	1	

27. Draw the 4-input K-map and use it to simplify the Boolean expression in Step 25. Write the simplified Boolean expression.

28. Click the "Truth Table to Simplified Boolean Expression" button on the logic converter.

> NOTE: This step can't be performed in a hardwired laboratory environment.

Question: How does your simplified Boolean expression in Step 27 compare with the Boolean expression on the logic converter?

29. Draw the simplified logic circuit from the simplified Boolean expression in Step 27.

30. Click the "Boolean Expression to Logic Circuit" button on the logic converter. Double click the logic converter in the upper left hand corner to reduce the enlargement.

> NOTE: A logic converter is not available in a hardwired laboratory. If you are performing this experiment in a hardwired laboratory, build the simplified logic circuit in Step 29, use logic switches for inputs A, B, C, and D, and use a logic probe light to monitor output (X). Measure the outputs for each input combination to complete the truth table (Table 8-10) for the simplified logic circuit.

Table 8-10 Truth Table

A	B	C	D	X
0	0	0	0	
0	0	0	1	
0	0	1	0	
0	0	1	1	
0	1	0	0	
0	1	0	1	
0	1	1	0	
0	1	1	1	
1	0	0	0	
1	0	0	1	
1	0	1	0	
1	0	1	1	
1	1	0	0	
1	1	0	1	
1	1	1	0	
1	1	1	1	

Question: How does your simplified logic circuit in Step 29 compare with the logic circuit in red on the screen (or how does the truth table for the simplified logic circuit compare with the truth table for the original circuit)?

31. Pull down the File menu and open FIG8-6. Write the Boolean expression for the logic circuit in Figure 8-6.

32. Click the "Circuit to Truth Table" button on the logic converter.

NOTE: A logic converter is not available in a hardwired laboratory. If you are performing this experiment in a hardwired laboratory, build the logic circuit in Figure 8-6, use logic switches for inputs A, B, C, and D, and use a logic probe light to monitor output (X). Measure the outputs for each input combination to complete the truth table (Table 8-11) for the logic circuit.

Table 8-11 Truth Table

A	B	C	D	X
0	0	0	0	
0	0	0	1	
0	0	1	0	
0	0	1	1	
0	1	0	0	
0	1	0	1	
0	1	1	0	
0	1	1	1	
1	0	0	0	
1	0	0	1	
1	0	1	0	
1	0	1	1	
1	1	0	0	
1	1	0	1	
1	1	1	0	
1	1	1	1	

33. Draw the 4-input K-map and use it to simplify the Boolean expression in Step 31. Write the simplified Boolean expression.

34. Click the "Truth Table to Simplified Boolean Expression" button on the logic converter.

> NOTE: This step can't be performed in a hardwired laboratory environment.

Question: How does your simplified Boolean expression in Step 33 compare with the Boolean expression on the logic converter?

35. Draw the simplified logic circuit from the simplified Boolean expression in Step 33.

36. Click the "Boolean Expression to Logic Circuit" button on the logic converter. Double click the logic converter in the upper left hand corner to reduce the enlargement.

> NOTE: A logic converter is not available in a hardwired laboratory. If you are performing this experiment in a hardwired laboratory, build the simplified logic circuit in Step 35, use logic switches for inputs A, B, C, and D, and use a logic probe light to monitor output (X). Measure the outputs for each input combination to complete the truth table (Table 8-12) for the simplified logic circuit.

Table 8-12 Truth Table

A	B	C	D	X
0	0	0	0	
0	0	0	1	
0	0	1	0	
0	0	1	1	
0	1	0	0	
0	1	0	1	
0	1	1	0	
0	1	1	1	
1	0	0	0	
1	0	0	1	
1	0	1	0	
1	0	1	1	
1	1	0	0	
1	1	0	1	
1	1	1	0	
1	1	1	1	

Question: How does your simplified logic circuit in Step 35 compare with the logic circuit in red on the screen (or how does the truth table for the simplified logic circuit compare with the truth table for the original circuit)?

Name ____________________
Date ____________________

Designing Combinational Logic Circuits

Objectives:

1. Obtain experience developing truth tables from problem statements.
2. Obtain experience writing sum-of-product (AND-OR) logic expressions from truth tables.
3. Obtain experience designing the simplest NAND-NAND combinational logic circuits from logic expressions.
4. Test combinational logic circuit designs for conformity with the original problem statements or truth tables.

Materials:

One 5 V dc power supply
Four logic switches
One logic probe light
Two-input NAND gates (1-7400 IC)
Three-input NAND gates (1-7410 IC)

Preparation:

In this experiment, you will design several combinational logic circuits from problem statements. First, you will develop a truth table from the problem statement, write the AND-OR logic expression from the truth table, simplify the logic expression using a K-map, and draw the simplified AND-OR logic circuit from the simplified logic expression. Then you will convert the AND-OR logic circuit to NAND-NAND logic and build and test the NAND-NAND logic circuit. If you build the NAND-NAND logic circuit on the computer screen using Electronics Workbench, you will use the logic converter to determine if the truth table for your circuit design matches the original truth table from the problem statement. If you build the NAND-NAND logic circuit in a hardwired laboratory, you will determine if the truth table for your circuit design matches the original truth table from the problem statement using logic switches to control the logic inputs and a logic probe light to monitor the output.

A truth table lists the outputs for all of the possible input combinations for a logic circuit. A sum-of-products (AND-OR) Boolean expression can be derived from a truth table by determining the input combinations required to produce a binary "1" output and representing each of these input combinations as an AND operation. The Boolean expression is determined by ORing all of the AND operations, resulting in an AND-OR logic expression.

Review the Preparation section of Experiment 8 for information on using K-maps to simplify logic expressions. Review the Preparation section of Experiment 6 for information on how to convert AND-OR logic to NAND-NAND logic.

Procedure:

1. Develop the truth table for a 2-input combinational logic circuit whose output is high only when one of the inputs is high.

2. Write the sum-of-products (AND-OR) Boolean expression for the truth table in Step 1.

3. Using a K-map, simplify the Boolean expression in Step 2, if possible.

Question: Were you able to simplify the Boolean expression? Was the original expression already in its simplest form?

4. Draw the simplest AND-OR logic circuit from the simplest Boolean expression in Step 3.

5. Draw the equivalent NAND-NAND logic circuit for the AND-OR logic circuit in Step 4. Use NAND gates to implement INVERTERS.

6. Construct the NAND-NAND logic circuit drawn in Step 5. If you construct the logic circuit on the computer screen using Electronics Workbench, use the logic converter to compare the truth table of your simplest NAND-NAND logic circuit design with the original truth table from the problem statement in Step 1. If you construct the logic circuit in a hardwired laboratory, use logic switches for the logic circuit inputs and use a logic probe light to monitor the output for each input combination to determine if your circuit truth table matches the truth table from the problem statement.

Question: How did the truth table for your simplified NAND-NAND logic circuit compare with the truth table from the problem statement in Step 1? Did your logic circuit design satisfy the conditions of the stated problem?

7. Develop the truth table for a 3-input combinational logic circuit whose output is high when a majority of the inputs are high.

8. Write the sum-of-products (AND-OR) Boolean expression for the truth table in Step 7.

9. Using a K-map, simplify the Boolean expression in Step 8, if possible.

Question: Were you able to simplify the Boolean expression? Was the original expression already in its simplest form?

10. Draw the simplest AND-OR logic circuit from the simplest Boolean expression in Step 9.

11. Draw the equivalent NAND-NAND logic circuit for the AND-OR logic circuit in Step 10. Use NAND gates to implement INVERTERS.

12. Construct the NAND-NAND logic circuit drawn in Step 11. If you construct the logic circuit on the computer screen using Electronics Workbench, use the logic converter to compare the truth table of your simplest NAND-NAND logic circuit design with the original truth table from the problem statement in Step 7. If you construct the logic circuit in a hardwired laboratory, use logic switches for the logic circuit inputs and use a logic probe light to monitor the output for each input combination to determine if your circuit truth table matches the truth table from the problem statement.

Question: How did the truth table for your simplified NAND-NAND logic circuit compare with the truth table from the problem statement in Step 7? Did your logic circuit design satisfy the conditions of the stated problem?

13. Develop the truth table for a 3-input combinational logic circuit whose output is the inverse of input A when inputs B and C are equal, and whose output is high when inputs B and C are different.

14. Write the sum-of-products (AND-OR) Boolean expression for the truth table in Step 13.

15. Using a K-map, simplify the Boolean expression in Step 14, if possible.

Question: Were you able to simplify the Boolean expression? Was the original expression already in its simplest form?

16. Draw the simplest AND-OR logic circuit from the simplest Boolean expression in Step 15.

17. Draw the equivalent NAND-NAND logic circuit for the AND-OR logic circuit in Step 16. Use NAND gates to implement INVERTERS.

18. Construct the NAND-NAND logic circuit drawn in Step 17. If you construct the logic circuit on the computer screen using Electronics Workbench, use the logic converter to compare the truth table of your simplest NAND-NAND logic circuit design with the original truth table from the problem statement in Step 13. If you construct the logic circuit in a hardwired laboratory, use logic switches for the logic circuit inputs and use a logic probe light to monitor the output for each input combination to determine if your circuit truth table matches the truth table from the problem statement.

Question: How did the truth table for your simplified NAND-NAND logic circuit compare with the truth table from the problem statement in Step 13? Did your logic circuit design satisfy the conditions of the stated problem?

19. **Develop the truth table for a 4-input combinational logic circuit that will produce a high output when the 4-bit binary input is greater than 9, and will produce a low output when the 4-bit binary input is 9 or less.**

20. Write the sum-of-products (AND-OR) Boolean expression for the truth table in Step 19.

21. Using a K-map, simplify the Boolean expression in Step 20, if possible.

Question: Were you able to simplify the Boolean expression? Was the original expression already in its simplest form?

22. Draw the simplest AND-OR logic circuit from the simplest Boolean expression in Step 21.

23. Draw the equivalent NAND-NAND logic circuit for the AND-OR logic circuit in Step 22. Use NAND gates to implement INVERTERS.

24. Construct the NAND-NAND logic circuit drawn in Step 23. If you construct the logic circuit on the computer screen using Electronics Workbench, use the logic converter to compare the truth table of your simplest NAND-NAND logic circuit design with the original truth table from the problem statement in Step 19. If you construct the logic circuit in a hardwired laboratory, use logic switches for the logic circuit inputs and use a logic probe light to monitor the output for each input combination to determine if your circuit truth table matches the truth table from the problem statement.

Question: How did the truth table for your simplified NAND-NAND logic circuit compare with the truth table from the problem statement in Step 19? Did your logic circuit design satisfy the conditions of the stated problem?

25. Develop the truth table for a 4-input combinational logic circuit that will output a high whenever the 4-bit binary input represents an even number and output a low whenever the 4-bit binary input represents an odd number.

26. Write the sum-of-products (AND-OR) Boolean expression for the truth table in Step 25.

27. Using a K-map, simplify the Boolean expression in Step 26, if possible.

Question: Were you able to simplify the Boolean expression? Was the original expression already in its simplest form?

28. Draw the simplest AND-OR logic circuit from the simplest Boolean expression in Step 27.

29. Draw the equivalent NAND-NAND logic circuit for the AND-OR logic circuit in Step 28. Use NAND gates to implement INVERTERS.

30. Construct the NAND-NAND logic circuit drawn in Step 29. If you construct the logic circuit on the computer screen using Electronics Workbench, use the logic converter to compare the truth table of your simplest NAND-NAND logic circuit design with the original truth table from the problem statement in Step 25. If you construct the logic circuit in a hardwired laboratory, use logic switches for the logic circuit inputs and use a logic probe light to monitor the output for each input combination to determine if your circuit truth table matches the truth table from the problem statement.

Question: How did the truth table for your simplified NAND-NAND logic circuit compare with the truth table from the problem statement in Step 25? Did your logic circuit design satisfy the conditions of the stated problem?

Name ______________________

Date ______________________

Troubleshooting Combinational Logic Circuits

Objectives:

1. Determine the defective logic gates for various combinational logic circuits by monitoring the logic levels at circuit test points.

Materials:

This experiment can only be performed on Electronics Workbench using the circuits disk provided with this manual.

Preparation:

In order to perform this experiment effectively, you must first complete Experiments 1-6. Use the theory learned in those experiments to find the defective logic gates in the logic circuits in this experiment.

Determine the defective gate in each experiment by trying all possible logic network binary input combinations and noting the binary inputs and outputs on each logic gate. A logic gate is defective if it has an incorrect output for any combination of binary inputs, unless one of the inputs is connected to a defective logic gate with an open output. Remember that an open input on a 7400 series logic gate will behave as if there is a logical "one" on that open input terminal, even if that open input is being caused by the open output of another logic gate and the logic probe light is indicating a zero. For example, if one of the inputs of an OR gate that is not defective is connected to a logic gate with an open output, the OR gate will act as if there is a logical "one" on that input and a logical "one" on its output (logic probe light "on"), even if all of the OR gate inputs indicate logical "zero" (logic probe lights "off"). This will happen because an OR gate produces a logical "one" at the output when any input is at a logical "one" (or open). For this reason, make sure that you try all possible logic network binary input combinations before concluding which logic gate is defective.

Procedure:

Don't forget to read the Preparation section before attempting to determine the defective gates in the following logic circuits.

1. Pull down the File menu and open FIG10-1. Click the On-Off switch to run the analysis. Based on the logic levels at the test points for various logic network binary inputs, determine which logic gate is defective. To switch the logic switches, press the key on the computer keyboard that matches the letter label on the switch.

   Defective gate ________________

2. Pull down the File menu and open FIG10-2. Click the On-Off switch to run the analysis. Based on the logic levels at the test points for various logic network binary inputs, determine which logic gate is defective. To switch the logic switches, press the key on the computer keyboard that matches the letter label on the switch.

   Defective gate ________________

3. Pull down the File menu and open FIG10-3. Click the On-Off switch to run the analysis. Based on the logic levels at the test points for various logic network binary inputs, determine which logic gate is defective. To switch the logic switches, press the key on the computer keyboard that matches the letter label on the switch.

   Defective gate ________________

4. Pull down the File menu and open FIG10-4. Click the On-Off switch to run the analysis. Based on the logic levels at the test points for various logic network binary inputs, determine which logic gate is defective. To switch the logic switches, press the key on the computer keyboard that matches the letter label on the switch.

   Defective gate ________________

5. Pull down the File menu and open FIG10-5. Click the On-Off switch to run the analysis. Based on the logic levels at the test points for various logic network binary inputs, determine which logic gate is defective. To switch the logic switches, press the key on the computer keyboard that matches the letter label on the switch.

   Defective gate ________________

6. Pull down the File menu and open FIG10-6. Click the On-Off switch to run the analysis. Based on the logic levels at the test points for various logic network binary inputs, determine which logic gate is defective. To switch the logic switches, press the key on the computer keyboard that matches the letter label on the switch.

   Defective gate ________________

7. Pull down the File menu and open FIG10-7. Click the On-Off switch to run the analysis. Based on the logic levels at the test points for various logic network binary inputs, determine which logic gate is defective. To switch the logic switches, press the key on the computer keyboard that matches the letter label on the switch.

   Defective gate ________________

8. Pull down the File menu and open FIG10-8. Click the On-Off switch to run the analysis. Based on the logic levels at the test points for various logic network binary inputs, determine which logic gate is defective. To switch the logic switches, press the key on the computer keyboard that matches the letter label on the switch.

   Defective gate ________________

9. Pull down the File menu and open FIG10-9. Click the On-Off switch to run the analysis. Based on the logic levels at the test points for various logic network binary inputs, determine which logic gate is defective. To switch the logic switches, press the key on the computer keyboard that matches the letter label on the switch.

   Defective gate ________________

10. Pull down the File menu and open FIG10-10. Click the On-Off switch to run the analysis. Based on the logic levels at the test points for various logic network binary inputs, determine which logic gate is defective. To switch the logic switches, press the key on the computer keyboard that matches the letter label on the switch.

    Defective gate ________________

11. Pull down the File menu and open FIG10-11. Click the On-Off switch to run the analysis. Based on the logic levels at the test points for various logic network binary inputs, determine which logic gate is defective. To switch the logic switches, press the key on the computer keyboard that matches the letter label on the switch.

    Defective gate ________________

12. Pull down the File menu and open FIG10-12. Click the On-Off switch to run the analysis. Based on the logic levels at the test points for various logic network binary inputs, determine which logic gate is defective. To switch the logic switches, press the key on the computer keyboard that matches the letter label on the switch.

    Defective gate ________________

PART

Arithmetic Logic Circuits

The experiments in Part II involve the study of arithmetic logic circuits. You will study XOR and XNOR gates, and then determine how they are used to build half-adders, full-adders, binary adders, BCD adders, parity generators and checkers, and magnitude comparators. In the final experiment in Part II you will solve some troubleshooting problems in arithmetic logic circuits.

The circuits for the experiments in Part II can be found on the enclosed disk in the PART2 subdirectory.

Name ____________________

Date ____________________

Logic Gates: XOR and XNOR

Objectives:

1. Complete the truth table for an XOR gate and compare it with the OR gate truth table.
2. Plot the XOR gate pulse response timing diagram.
3. Complete the truth table for an XNOR gate and compare it with the XOR gate truth table.
4. Plot the XNOR gate pulse response timing diagram.
5. Demonstrate the use of an XOR gate as a controlled inverter.
6. Design the simplest AND-OR logic network that satisfies the XOR truth table.
7. Design the simplest AND-OR logic network that satisfies the XNOR truth table.

Materials:

One 5 V dc voltage supply
Two logic switches
Three logic probe lights
One logic analyzer
Two pulse generators
One XOR gate (1-7486 IC)
INVERTERS (1-7404 IC)
Two-input AND gates (1-7408 IC)
Two-input OR gates (1-7432 IC)
Resistors—1 kΩ (2), 10 kΩ

Preparation:

A truth table lists the outputs for all of the possible input combinations for a logic gate or logic network. See the results in Experiment 3 for the OR gate truth table.

The XOR gate output will be high (1) when only one of the inputs is high (1). The XOR gate output will be low (0) when both of the inputs are low (0) or high (1). A circuit for studying the XOR gate is shown in Figure 11-1. The circuit for plotting the XOR gate pulse response timing diagram is shown in Figure 11-2.

An XNOR gate is usually wired using an XOR gate with an INVERTER on the output. The output will be high (1) when both of the inputs are low (0) or high (1). The output will be low (0) when only one of the inputs is high (1). A circuit for studying the XNOR gate is shown in Figure 11-3. The circuit for plotting the XNOR gate pulse response timing diagram is shown in Figure 11-4.

If one of the inputs of an XOR gate is connected to a logic switch, as shown in Figure 11-5, it can be used as a controlled inverter. The XOR gate output (Y) will be equal to the other XOR gate input (A) when the logic switch is low (0), and the XOR gate output (Y) will be the inverse of the other XOR gate input (A) when the logic switch is high (1). This can be demonstrated from the XOR gate truth table.

Complete Experiment 9 before designing the XOR and XNOR logic circuits in Steps 7–12. If you have already completed this experiment, you may want to review the Preparation section.

Figure 11-1 XOR Gate

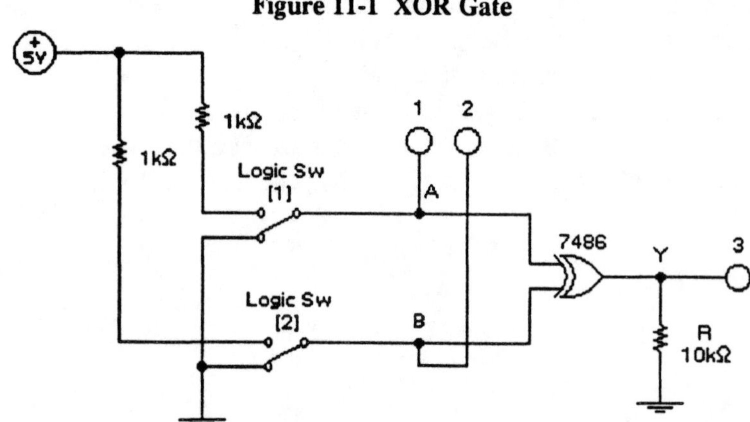

Figure 11-2a XOR Gate Pulse Response (EWB Verson 4)

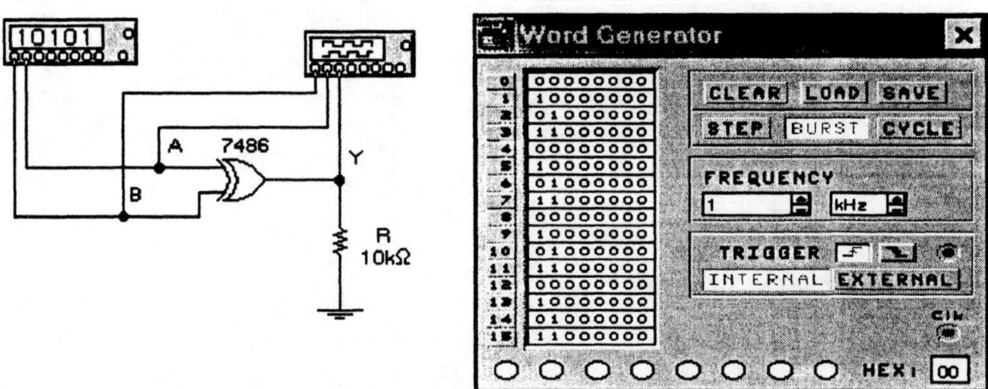

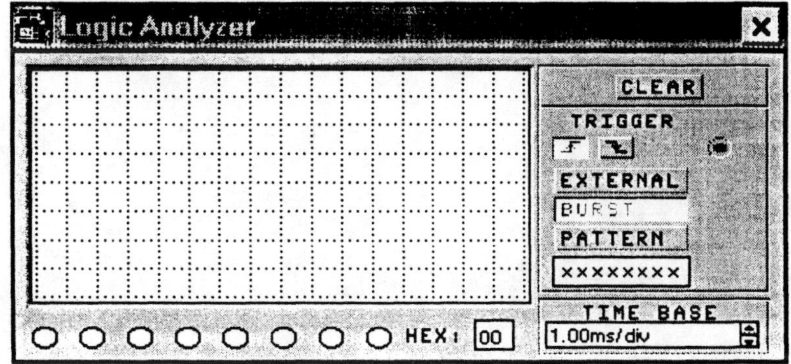

Figure 11-2b XOR Gate Pulse Response (EWB Version 5)

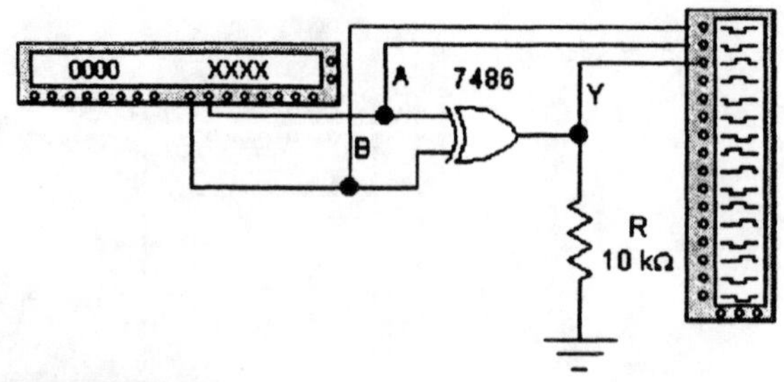

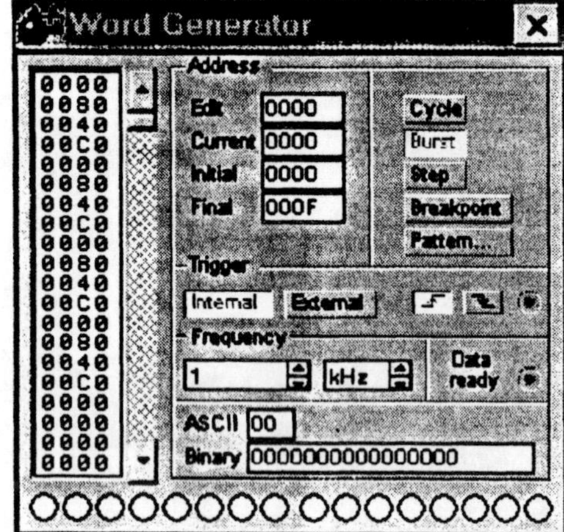

Logic Analyzer Settings
Clocks per division ---------------- 8

Clock Setup dialog box
Clock edge --------------- positive
Clock mode -------------- Internal
Internal clock rate ------ 10 kHz
Clock qualifier ----------- x
Pre-trigger samples --- 100
Post-trigger samples -- 1000
Threshold voltage (V) 2

Trigger Patterns dialog box
A ------------------------------ xxxxxxxxxxxxxxxx
Trigger combinations -- A
Trigger qualifier ---------- x

Figure 11-3 XNOR Gate

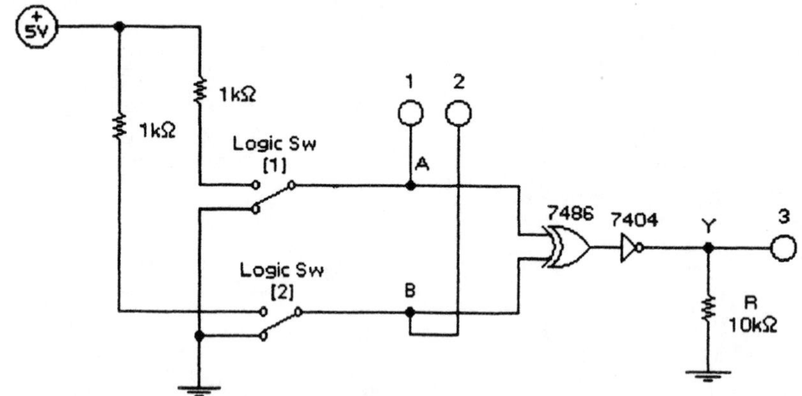

Figure 11-4a XNOR Gate Pulse Response (EWB Version 4)

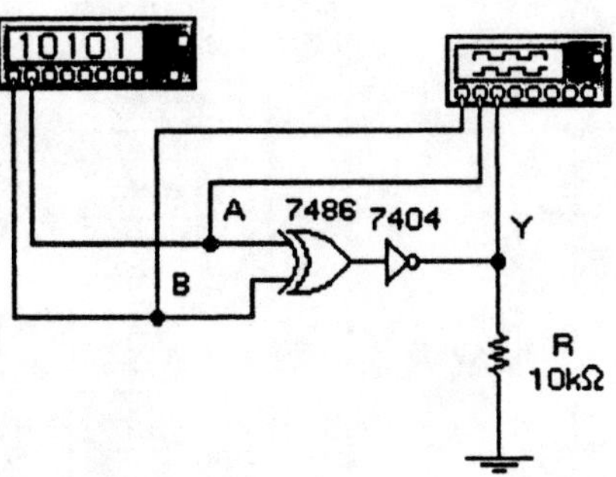

Figure 11-4b XNOR Gate Pulse Response (EWB Version 5)

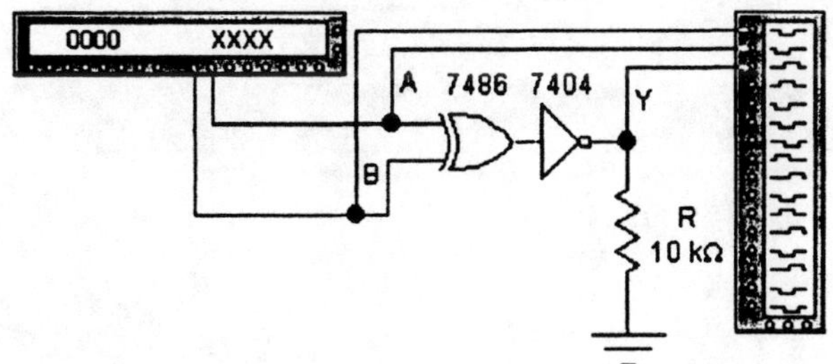

Figure 11-5a XOR Gate Controlled Inverter (EWB Version 4)

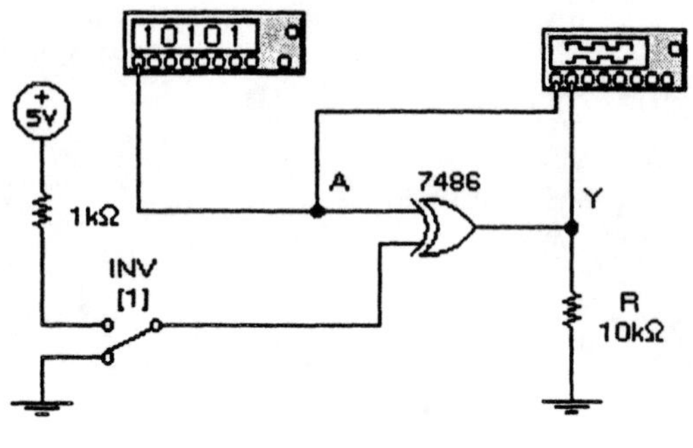

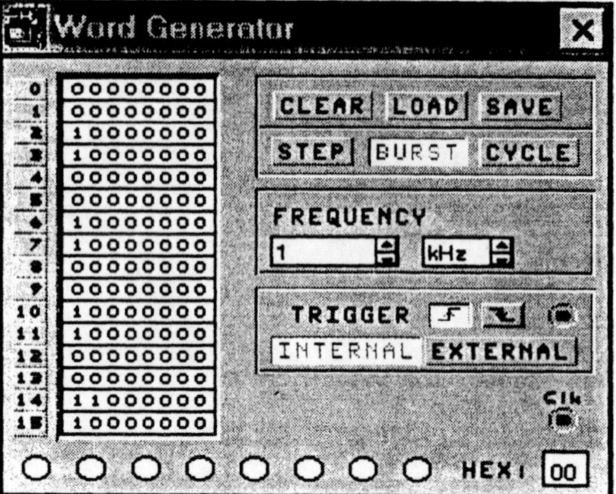

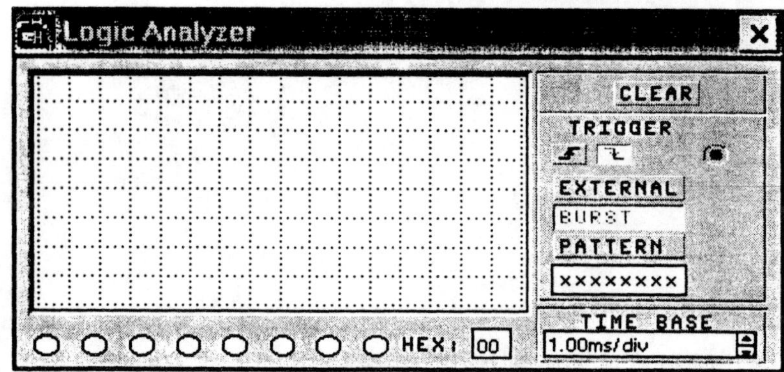

Figure 11-5b XOR Gate Controlled Inverter (EWB Version 5)

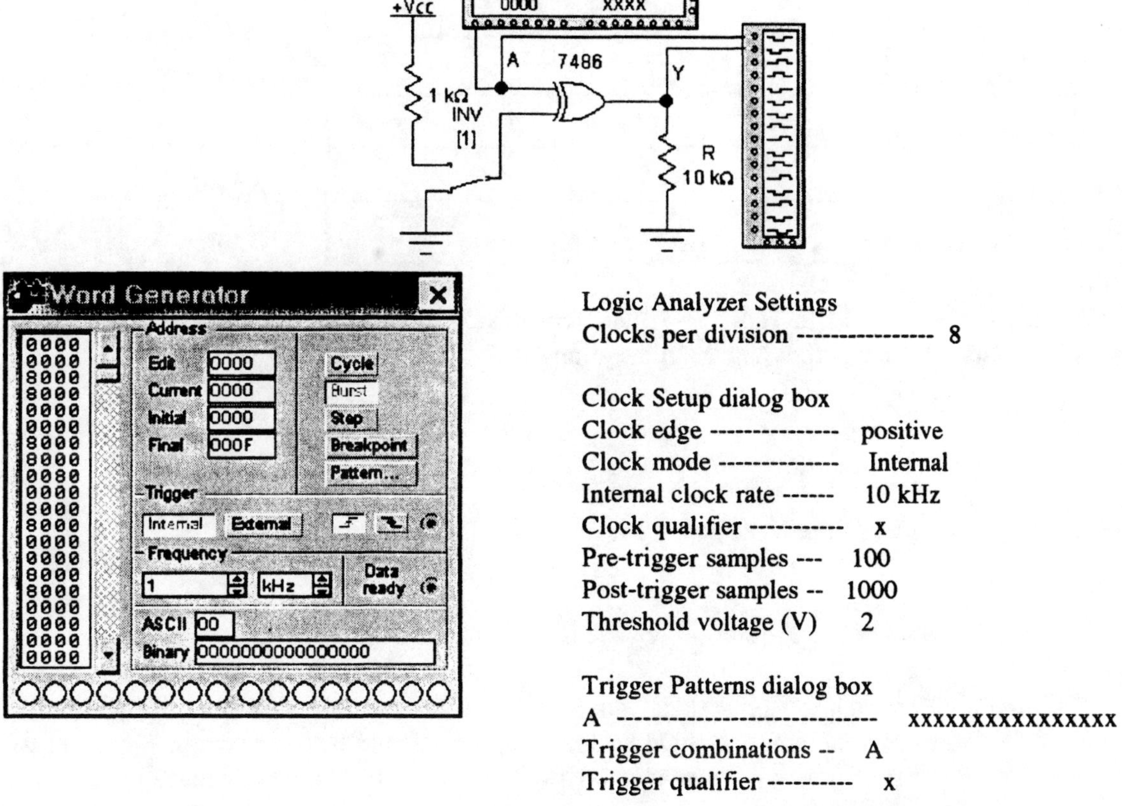

Procedure:

1. Pull down the File menu and open FIG11-1. Notice that Logic Switches 1 and 2 are placing binary "zeros" (ground equals 0 V) on the two XOR gate inputs (A and B). Resistor R represents the load on the XOR gate output. Click the On-Off switch to run the analysis. By switching the logic switches to the appropriate positions, complete the XOR gate truth table (Table 11-1). You can switch the logic switches by pressing the number on the keyboard corresponding to the number label on the switch.

Table 11-1 XOR Gate Truth Table

A	B	Y
0	0	
0	1	
1	0	
1	1	

Questions: Based on the results in Table 11-1, what conclusion can you draw about the relationship between the XOR gate binary output and the binary inputs?

What is the difference between the XOR gate truth table and the OR gate truth table?

2. Click the On-Off switch to stop the analysis run. Pull down the File menu and open FIG11-2. The word generator and logic analyzer settings should be as shown in Figure 11-2. Click the On-Off switch to run the analysis. Notice that the word generator has applied a pulse pattern to each XOR gate input, shown in red and green on the logic analyzer screen. The blue curve plot is the XOR gate output. Draw the XOR gate input and output curve plots in the space provided and label them.

NOTE: In a hardwired laboratory, use pulse generator outputs to apply square waves to the XOR gate inputs.

Question: Does the XOR gate output go high only when one of the inputs is high? Is this expected for an XOR gate?

3. Click the On-Off switch to stop the analysis run. Pull down the File menu and open FIG11-3. The XOR gate with an INVERTER on the output is equivalent to an XNOR gate. Notice that Logic Switches 1 and 2 are placing binary zeros (ground equals 0 V) on the two XNOR gate inputs (A and B). Resistor R represents the load on the XNOR gate output. Click the On-Off switch to run the analysis. By switching the logic switches to the appropriate positions, complete the XNOR gate truth table (Table 11-2).

Table 11-2 XNOR Gate Truth Table

A	B	Y
0	0	
0	1	
1	0	
1	1	

Questions: Based on the results in Table 11-2, what conclusion can you draw about the relationship between the XNOR gate binary output and the binary inputs?

What is the difference between the XNOR gate truth table and the XOR gate truth table? Explain.

4. Click the On-Off switch to stop the analysis run. Pull down the File menu and open FIG11-4. The word generator and logic analyzer settings should be as shown in Figure 11-2. Click the On-Off switch to run the analysis. Notice that the word generator has applied a pulse pattern to each XNOR gate input, shown in red and green on the logic analyzer screen. The blue curve plot is the XNOR gate output. Draw the XNOR gate input and output curve plots in the space provided and label them.

NOTE: In a hardwired laboratory, use pulse generator outputs to apply square waves to the XNOR gate inputs.

Question: Does the XNOR gate output go high only when both of the inputs are high or low? Is this expected for an XNOR gate?

5. Click the On-Off switch to stop the analysis run. Pull down the File menu and open FIG11-5. The INV switch should be down (0). The INV switch can be changed by pressing the "1" key on the keyboard. The word generator and logic analyzer settings should be as shown in Figure 11-5. Click the On-Off switch to run the analysis. Notice that the word generator has applied a pulse pattern to the XOR gate input (A), shown in red on the logic analyzer screen. The blue curve plot is the XOR gate output (Y). Draw the input (A) and output (Y) curve plots in the space provided and label them.

NOTE: In a hardwired laboratory, use a pulse generator in place of the word generator and an oscilloscope in place of the logic analyzer.

Question: What is the relationship between the input curve plot (A) and the output curve plot (Y)?

6. Click the On-Off switch to stop the analysis run. Press the "1" key to switch the INV switch to the up (1) position. Click the On-Off switch to run the analysis again. Draw the input (A) and output (Y) curve plots in the space provided and label them.

Questions: What is the relationship between the input curve plot (A) and the output curve plot (Y)?

What conclusion can you draw about the relationship between the output (Y) and the input (A) based on the position of the INV switch?

7. Click the On-Off switch to stop the analysis. Based on the XOR gate truth table (Table 11-1) write the simplest AND-OR logic equation that will satisfy the truth table.

8. Based on the logic equation in Step 7, draw the simplest AND-OR logic circuit that satisfies the equation.

9. Construct the AND-OR logic circuit drawn in Step 8. If you construct the logic circuit on the computer screen using Electronics Workbench, use the logic converter to compare the truth table of your simplest AND-OR logic circuit design with the original truth table (Table 11-1). If you construct the logic circuit in a hardwired laboratory, use logic switches for the logic circuit inputs and use a logic probe light to monitor the output. For each input combination, determine if your circuit truth table matches the XOR truth table in Table 11-1.

Questions: How did the truth table for your AND-OR logic circuit compare with the XOR truth table in Table 11-1?

Is your AND-OR logic circuit design equivalent to the XOR gate?

10. **Based on the XNOR gate truth table (Table 11-2) write the simplest AND-OR logic equation that will satisfy the truth table.**

11. **Based on the logic equation in Step 10, draw the simplest AND-OR logic circuit that satisfies the equation.**

12. Construct the AND-OR logic circuit drawn in Step 11. If you construct the logic circuit on the computer screen using Electronics Workbench, use the logic converter to compare the truth table of your simplest AND-OR logic circuit design with the original truth table (Table 11-2). If you construct the logic circuit in a hardwired laboratory, use logic switches for the logic circuit inputs and use a logic probe light to monitor the output. For each input combination, determine if your circuit truth table matches the XNOR truth table in Table 11-2.

Questions: How did the truth table for your AND-OR logic circuit compare with the XNOR truth table in Table 11-2?

Is your AND-OR logic circuit design equivalent to the XNOR gate?

Name ____________________

Date ____________________

Arithmetic Circuits

Objectives:

1. Design a binary half-adder using an XOR gate and an AND gate.
2. Design a binary full-adder using two half-adders.
3. Design a binary full-adder using AND-OR logic.

Materials:

One 5 V dc power supply
Three logic switches
Two logic probe lights
XOR gates (1-7486 IC)
Two-input AND gates (1-7408 IC)
Three-input AND gates (2-7411 ICs)
Two-input OR gates (1-7432 IC)
INVERTERS (1-7404 IC)
1 kΩ resistors

Preparation:

Before designing the half-adder in Steps 1–5, make sure you complete Experiment 11 on XOR gates. If this experiment has been completed, review the Preparation section. You should also review Experiment 2 on AND gates.

A half-adder adds two binary input bits (A and B) and produces a sum output (S) and a carry output (C). The carry output (C) will go to a binary one (1) when the two input bits (A and B) are both one (1), and the sum output (S) will go to zero. The truth table for the half-adder is shown in Table 12-1.

A full-adder adds two binary input bits (A and B) and a carry input bit (Cin) and produces a sum output (S) and a carry output (Co). The carry output (Co) will go to a binary one (1) when two or more of the input bits are binary one (1). You can build a binary full-adder from two binary half-adders by connecting the sum output of the first half-adder to one of the inputs of the second half-adder, and connecting the carry input (Cin) to the other input of the second half-adder. This will produce the sum of the three input bits. If either half-adder produces a carry output, then the full-adder should produce a carry output (Co). Therefore, you will need to connect the carry output of each half-adder to a two-input OR gate. The output of the two-input OR gate is the full-adder carry output (Co). A truth table for a binary full-adder is shown in Table 12-2.

Before designing the full-adder using AND-OR logic in Steps 11-18, review Experiment 9 on designing combinational logic circuits and Experiment 8 on Karnaugh maps.

Table 12-1 Half-Adder Truth Table

A	B	C	S
0	0	0	0
0	1	0	1
1	0	0	1
1	1	1	0

Table 12-2 Full-Adder Truth Table

A	B	Cin	Co	S
0	0	0	0	0
0	0	1	0	1
0	1	0	0	1
0	1	1	1	0
1	0	0	0	1
1	0	1	1	0
1	1	0	1	0
1	1	1	1	1

Procedure:

1. The truth table for a binary half-adder is shown in Table 12-1. From the truth table, design a logic network to produce the half-adder sum (S) output using an XOR gate, and draw the network in the space provided.

2. From the half-adder truth table (Table 12-1), design a logic network to produce the carry (C) output using an AND gate and draw the network in the space provided.

3. Combine the two logic networks designed in Steps 1 and 2 and draw the complete network for the half-adder in the space provided. Use the same inputs (A and B) for both logic networks.

4. Wire the half-adder logic network drawn in Step 3 using logic switches for inputs A and B. (See the previous experiments using logic switches.) Use logic probe lights to monitor outputs S and C. You can wire the network on the computer using Electronics Workbench or in a hardwired laboratory. If you wire the circuit in a hardwired lab, don't forget to connect +5 V and ground to each IC chip.

5. Click the On-Off switch to run the analysis or turn on the power. Compare your sum (S) and carry (C) outputs with the half-adder truth table (Table 12-1) for all combinations of the inputs (A and B).

Question: Did your half-adder truth table match the truth table in Table 12-1? If not, why?

6. After your half-adder is working properly, save it on your own personal disk. (You can't save any circuits on the disk provided with this manual because it is write protected.)

NOTE: If the circuit was hardwired, save the circuit.

7. A full-adder should add input bits A and B plus a carry input bit (Cin) and produce a sum output (S) and a carry output (Co). Design a logic network that will produce the sum and carry outputs for inputs A, B, and Cin using two half-adders. Draw your logic network in the space provided.

8. Wire the full-adder drawn in Step 7 by adding a second half-adder to the half-adder wired in Step 4. Don't forget to add an additional logic switch for the carry input bit (Cin).

9. Click the On-Off switch to run the analysis or turn on the power. Compare your sum (S) and carry (Co) outputs with the full-adder truth table (Table 12-2) for all combinations of the inputs (A, B, and Cin).

Question: Did your full-adder truth table match the truth table in Table 12-2? If not, why?

10. After your full-adder is working properly, save it on your own personal disk. (You can't save any circuits on the disk provided with this manual because it is write protected.)

11. Based on the full-adder truth table (Table 12-2), develop the simplest AND-OR logic equation for the sum output (S). Use a K-map to simplify the equation, if possible.

12. Based on the full-adder truth table (Table 12-2), develop the simplest AND-OR logic equation for the carry output (Co). Use a K-map to simplify the equation, if possible.

13. Draw the AND-OR logic circuit to satisfy the logic equation developed for the sum output (S) in Step 11.

14. Draw the AND-OR logic circuit to satisfy the logic equation developed for the carry output (Co) in Step 12.

15. Wire the AND-OR logic circuit for the sum (S) output drawn in Step 13. Use three logic switches for inputs A, B, and Cin. Use a logic probe light to monitor the sum output (S). You can wire the network on the computer using Electronics Workbench or in a hardwired laboratory. If you wire the circuit in a hardwired lab, don't forget to connect +5 V and ground to each IC chip.

16. Wire the AND-OR logic circuit for the carry output (Co) drawn in Step 14. Use the same three logic switches used in Step 15 for inputs A, B, and Cin. Use a logic probe light to monitor the carry output (Co).

17. Click the On-Off switch to run the analysis or turn on the power. Compare your sum (S) and carry (Co) outputs with the full-adder truth table (Table 12-2) for all combinations of the inputs (A, B, and Cin).

Question: Did your full-adder truth table match the truth table in Table 12-2? If not, why?

18. After your full-adder is working properly, save it on your own personal disk. (You can't save any circuits on the disk provided with this manual because it is write protected.)

Name ____________________
Date ____________________

Parallel Binary Adder

Objectives:

1. Investigate the operation of a four-bit binary adder.
2. Obtain experience adding binary numbers.
3. Investigate the operation of a four-bit 2's complement adder/subtractor.
4. Obtain experience adding and subtracting binary numbers in the 2's complement number system.

Materials:

One 5 V dc power supply
Nine logic switches
Thirteen logic probe lights
One four-bit binary parallel adder (1-7483 IC)
Four XOR gates (1-7486 IC)

Preparation:

A four-bit binary adder consists of four full adders wired as shown in Figure 13-1. Notice that the carry output (Co) of each full adder is wired to the carry input (Ci) of the next full adder, from right-to-left. Therefore, each full adder will add one bit of the A number to one bit of the B number plus any carry input from the previous full adder, as shown in the following example.

```
      C2 C1 C0 0
      A3 A2 A1 A0
      B3 B2 B1 B0
      -----------
Co    S3 S2 S1 S0
```

Notice that binary zero (0) is being applied to the Ci terminal of the first full adder (binary adder carry input), adding 0 to the answer (S). If a binary one (1) were applied to the binary adder carry input, 1 would be added to the answer (S).

If two four-bit binary adders are connected together to produce an eight-bit adder, the carry output of the first four-bit binary adder on the right would be connected to the carry input (Ci) of the second four-bit binary adder on the left.

If this experiment is performed in a hardwired laboratory, a 7483 four-bit binary adder should be wired in place of the four full adders, as shown in Figure 13-2. The logic circuit inside the 7483 is similar to the logic shown in Figure 13-1. Pin 1 on the 7483 is located in the lower left hand corner of the chip, as designated by the dot in Figure 13-2. See Appendix A for the pin diagram of the 7483 IC chip.

In the 2's complement number system, the first bit of each binary number represents the sign bit and determines the sign of the binary number. A zero (0) sign bit represents a positive number and a one (1) sign bit represents a negative number. The 2's complement of a binary number is calculated by inverting each binary bit and then adding a binary one (1). Taking the 2's complement of a binary number reverses the sign of the number, making a positive number negative and a negative number positive. A negative binary number is represented by the 2's complement of its positive binary value. Therefore, adding a binary number to the 2's complement of another binary number is the equivalent of adding a positive number to a negative number, or the equivalent of subtracting one binary number from another binary number.

The four-bit binary adder/subtractor in Figure 13-3 operates in the 2's complement binary number system and consists of four full adders and four XOR gates. The four XOR gates are wired as four controlled inverters. (See Preparation section of Experiment 11.) They are controlled by the SUB switch in Figure 13-3. When the SUB switch is down (0), the XOR gates let each binary bit of the B number straight through to the full adders. This will cause the four-bit binary adder to add the A number to the B number. This is the ADD mode of the adder/subtractor. When the SUB switch is up (1), it places a binary one (1) on the carry input of the first full adder (four-bit adder carry input), and the XOR gates invert each bit of the B number. Inverting each bit of the B number and adding a binary one (1) is the same as taking the 2's complement of the B number. This will cause the four-bit adder to add the A number to the 2's complement of the B number, which is the equivalent of subtracting the B number from the A number. This is the SUBTRACT mode of the adder/subtractor.

If this adder/subtractor experiment is performed in a hardwired laboratory, a 7483 four-bit binary adder should be wired in place of the four full adders, as shown in Figure 13-4.

Because the four-bit binary numbers placed on the input of the adder/subtractor will be in the 2's complement number system, the range of numbers or answers cannot exceed decimal +7 (0111) or decimal -8 (1000). If this range of numbers is exceeded, an overflow of the number into the sign bit will occur. A larger number of bits is needed to represent a larger range of numbers.

In the subtraction mode, the adder/subtractor carry output becomes a borrow output and will go to a binary zero (0) when a larger number is subtracted from a smaller number. This represents a borrow from the next column. The carry (borrow) output will be a binary one (1) when a smaller number is subtracted from a larger number and there is no borrow.

Figure 13-1 Four-Bit Binary Adder

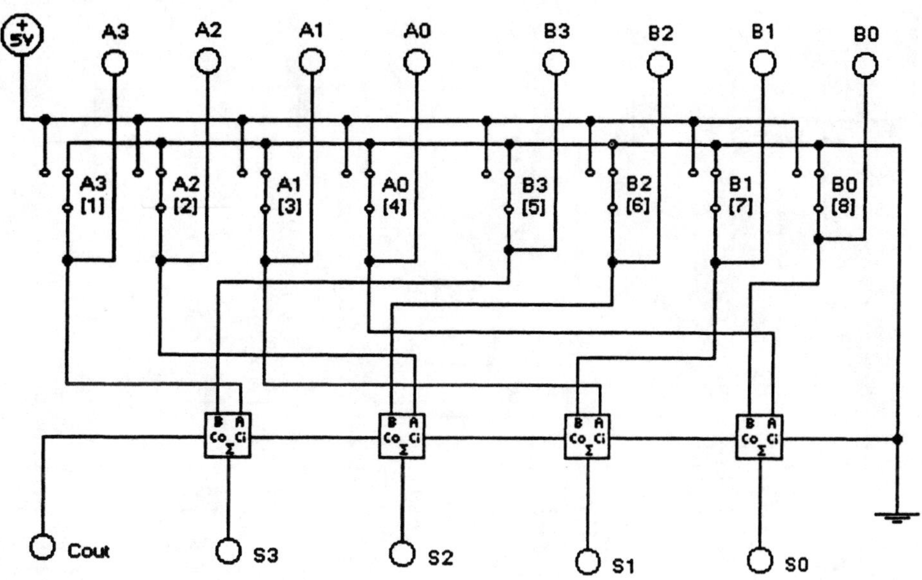

Figure 13-2 7483 Four-Bit Binary Adder

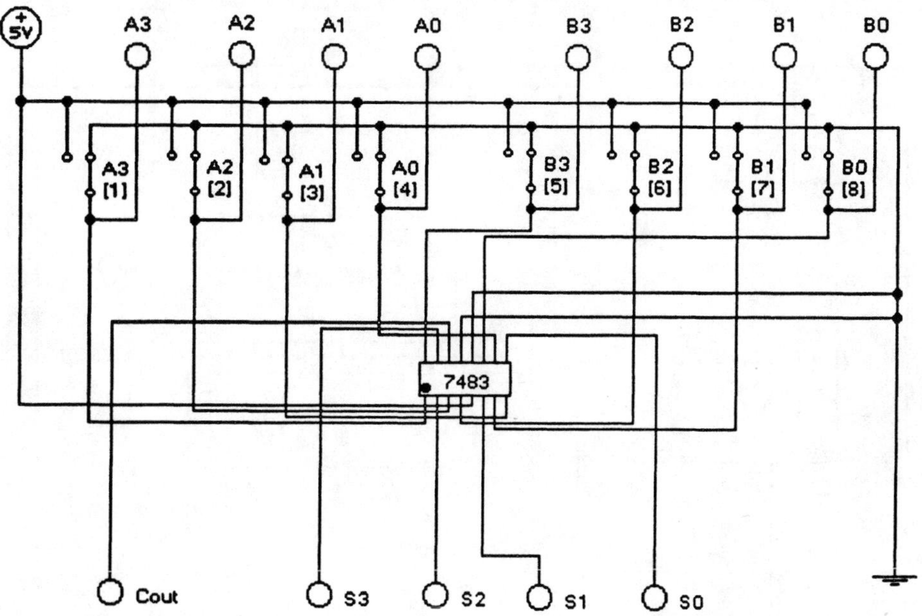

Figure 13-3 Four-Bit Binary Adder/Subtractor

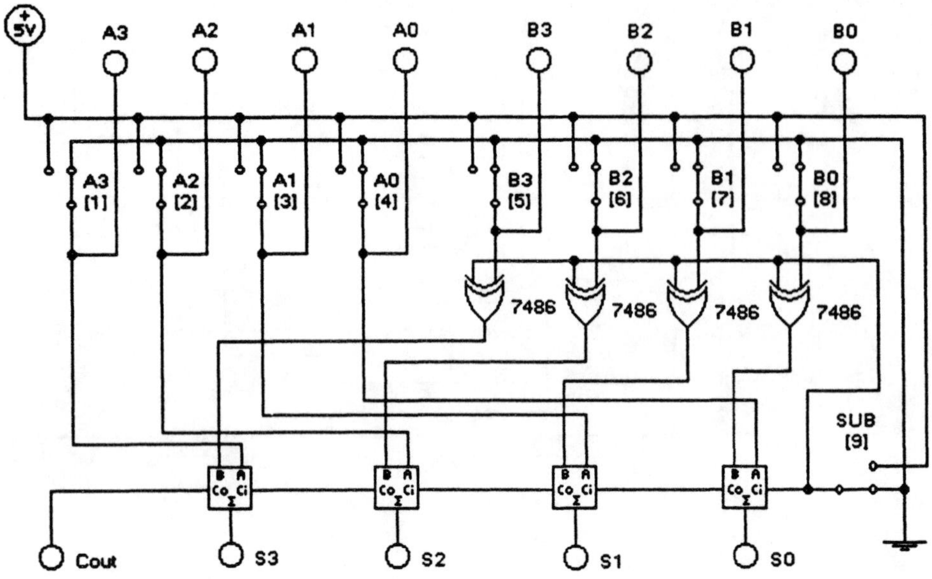

Figure 13-4 7483 Four-Bit Binary Adder/Subtractor

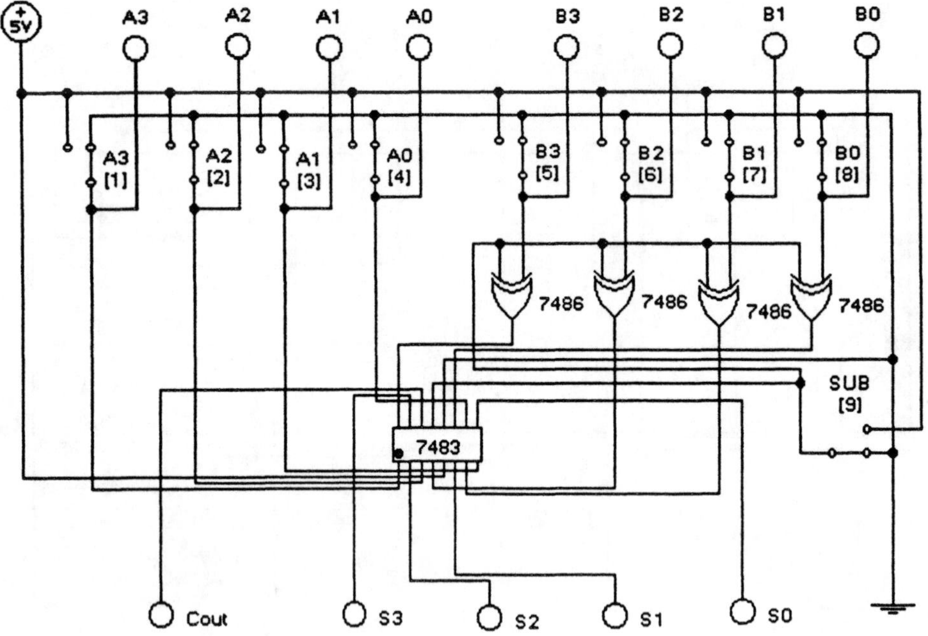

Procedure:

1. Pull down the File and open FIG13-1. Figure 13-1 shows a four-bit binary adder constructed from four full adders. See the Preparation section for more details. You will use this four-

bit binary adder to confirm the results when two four-bit binary numbers are added. Press number keys 1-8 to control the logic switches. If you are performing this experiment in a hardwired laboratory, wire the 7483 four-bit binary adder as shown in Figure 13-2.

2. Add binary 0111 to binary 0101 in the space provided and confirm your results using the four-bit binary adder in Figure 13-1 or Figure 13-2. Also show the decimal equivalent of the numbers and your answer.

Questions: Did your calculated answer match the answer on the adder?

Was the decimal equivalent of your answer correct?

3. Add binary 1111 to binary 1001 in the space provided and confirm your results using the four-bit binary adder in Figure 13-1 or Figure 13-2. Also show the decimal equivalent of the numbers and your answer.

Questions: Did your calculated answer match the answer on the adder?

Was the decimal equivalent of your answer correct?

4. Convert decimal 11 and decimal 10 to four-bit binary numbers and add them in the space provided. Confirm your results using the four-bit binary adder in Figure 13-1 or Figure 13-2.

Questions: Did your calculated answer match the answer on the adder?

Was the decimal equivalent of your answer correct?

5. **Convert decimal 13 and decimal 7 to four-bit binary numbers and add them in the space provided. Confirm your results using the four-bit binary adder in Figure 13-1 or Figure 13-2.**

Questions: Did your calculated answer match the answer on the adder?

Was the decimal equivalent of your answer correct?

6. **Click the On-Off switch to stop the analysis run. Pull down the File and open FIG13-3. Figure 13-3 shows a four-bit binary adder/subtractor constructed from four full adders and four XOR gates. Notice that the four XOR gates are wired as four controlled inverters, controlled by the SUB switch. When the SUB switch is down (0), the adder/subtractor will add the A number to the B number. When the SUB switch is up (1), the adder/subtractor will subtract the B number from the A number. Press the 9 key to control the SUB switch. See the Preparation section for more details. You will use this four-bit adder/subtractor to confirm the results when two four-bit binary numbers are added or subtracted *in the 2's complement number system*. Remember your numbers and answers cannot be outside the range of +7 (0111) to -8 (1000) or the number will overflow into the sign bit. If this adder/subtractor experiment is performed in a hardwired laboratory, a 7483 four-bit binary adder should be wired in place of the four full adders, as shown in Figure 13-4.**

7. Add 0110 to 0001 in the 2's complement number system and confirm your results using the four-bit binary adder/subtractor in Figure 13-3 or Figure 13-4. Also show the decimal equivalent of the numbers and your answer.

Questions: Did your calculated answer match the answer on the adder?

Was the decimal equivalent of your answer correct?

Was the sign bit correct?

8. **Subtract 0001 from 0110 in the 2's complement number system and confirm your results using the four-bit binary adder/subtractor in Figure 13-3 or Figure 13-4. Also show the decimal equivalent of the numbers and your answer.**

Questions: **Did your calculated answer match the answer on the adder?**

Was the decimal equivalent of your answer correct?

Was the sign bit correct?

9. **Subtract 0110 from 0001 in the 2's complement number system and confirm your results using the four-bit binary adder/subtractor in Figure 13-3 or Figure 13-4. Also show the decimal equivalent of the numbers and your answer.**

Questions: **Did your calculated answer match the answer on the adder?**

Was the decimal equivalent of your answer correct?

Was the sign bit correct?

10. **Add 1101 to 0011 in the 2's complement number system and confirm your results using the four-bit binary adder/subtractor in Figure 13-3 or Figure 13-4. Also show the decimal equivalent of the numbers and your answer.**

Questions: Did your calculated answer match the answer on the adder?

Was the decimal equivalent of your answer correct?

Was the sign bit correct?

11. Subtract 0011 from 0011 in the 2's complement number system and confirm your results using the four-bit binary adder/subtractor in Figure 13-3 or Figure 13-4. Also show the decimal equivalent of the numbers and your answer.

Questions: Did your calculated answer match the answer on the adder?

Was the decimal equivalent of your answer correct?

Was the sign bit correct?

12. Subtract 1101 from 0011 in the 2's complement number system and confirm your results using the four-bit binary adder/subtractor in Figure 13-3 or Figure 13-4. Also show the decimal equivalent of the numbers and your answer.

Questions: Did your calculated answer match the answer on the adder?

Was the decimal equivalent of your answer correct?

Was the sign bit correct?

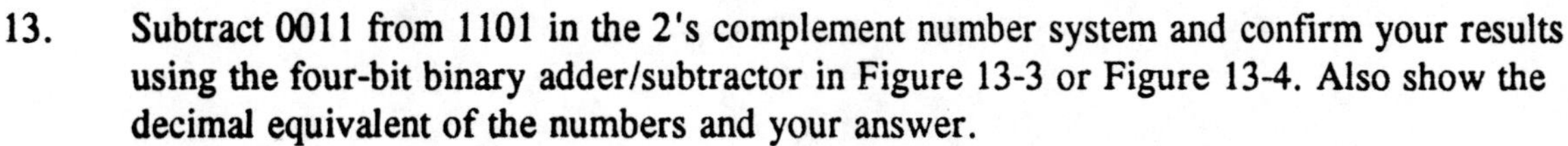

13. Subtract 0011 from 1101 in the 2's complement number system and confirm your results using the four-bit binary adder/subtractor in Figure 13-3 or Figure 13-4. Also show the decimal equivalent of the numbers and your answer.

Questions: Did your calculated answer match the answer on the adder?

Was the decimal equivalent of your answer correct?

Was the sign bit correct?

14. Convert decimal +6 and decimal -4 to four-bit 2's complement binary numbers and add them in the space provided. Confirm your results using the four-bit binary adder/subtractor in Figure 13-3 or Figure 13-4.

Questions: Did your calculated answer match the answer on the adder?

Was the decimal equivalent of your answer correct?

Was the sign bit correct?

15. Convert decimal -2 and decimal +5 to four-bit 2's complement binary numbers and subtract -2 from +5 in binary in the space provided. Confirm your results using the four-bit binary adder/subtractor in Figure 13-3 or Figure 13-4.

Questions: Did your calculated answer match the answer on the adder?

Was the decimal equivalent of your answer correct?

Was the sign bit correct?

Name ______________________________
Date ______________________________

BCD Adder

Objectives:

1. Investigate the operation of a BCD adder.
2. Obtain experience adding BCD numbers and correcting the sum.

Materials:

One 5 V dc power supply
Eight logic switches
Thirteen logic probe lights
Two four-bit binary parallel adders (2-7483 ICs)
One two-input AND gate (1-7408 IC)
Two two-input OR gates (1-7432 IC)

Preparation:

In the binary coded decimal (BCD) number system, each decimal number is represented by a 4-bit binary code from zero (0000) through nine (1001). The remaining six 4-bit numbers (1010-1111) are not part of the BCD code. Therefore, when BCD numbers are added, any answer that is larger than nine (1001) must be adjusted to the correct BCD code between zero (0000) and nine (1001), and a carry must be generated. The binary answer is corrected by adding binary six (0110) to the sum in order to skip the six invalid numbers. If the sum is nine (1001) or less, no adjustment is necessary.

A BCD adder is a 4-bit binary adder that adds two 4-bit binary numbers and determines if the sum is greater than nine (1001). If the sum is greater than nine (1001), binary six (0110) is added to the sum and a carry is generated. If the sum is not greater than nine (1001), binary zero (0000) is added and no carry is generated.

The BCD adder shown in Figure 14-1 includes two 4-bit binary adders and a logic network that detects when the first adder output is greater than nine (1001). Each of the two 4-bit binary adders consists of a bank of four full-adders. The first 4-bit binary adder (bank of four full-adders) will add the two 4-bit BCD inputs. The second 4-bit binary adder (bank of four full-adders) will add binary six (0110) or binary zero (0000) to the sum, depending on the output from the first adder.

The AND-OR logic network in Figure 14-1 will produce a binary one (1) output (X) when the output of the first adder is greater than nine (1001) or the first adder carry output (Co') is binary one (1). Otherwise, the logic network will produce a logic zero (0) output (X). This logic network output (X) will apply a binary one (1) or zero (0) to the BCD adder carry output (Cout) and to the second and third input bits of the second 4-bit binary adder. Because the first and fourth input bits of the second 4-bit binary adder are connected to ground (0), the second adder will add 0110 or 0000 to the sum from the first adder, depending on the output (X) of the logic network.

If this experiment is performed in a hardwired laboratory, two 7483 four-bit binary adders should be wired in place of the two banks of four full-adders, as shown in Figure 14-2. The logic circuit inside each 7483 is similar to each bank of four full-adders in Figure 14-1. Pin 1 on each 7483 is located in the lower left hand corner of the chip, as designated by the dot in Figure 14-2. See Appendix A for the pin diagram of the 7483 IC chip.

To add two 2-digit decimal numbers using the BCD code, two BCD adders must be cascaded. The carry output (Cout) of the first BCD adder must be connected to the carry input (Cin) of the second BCD adder. The number of BCD adders required to add numbers in the BCD code is determined by the number of digits in the decimal numbers to be added.

Figure 14-1 BCD Adder

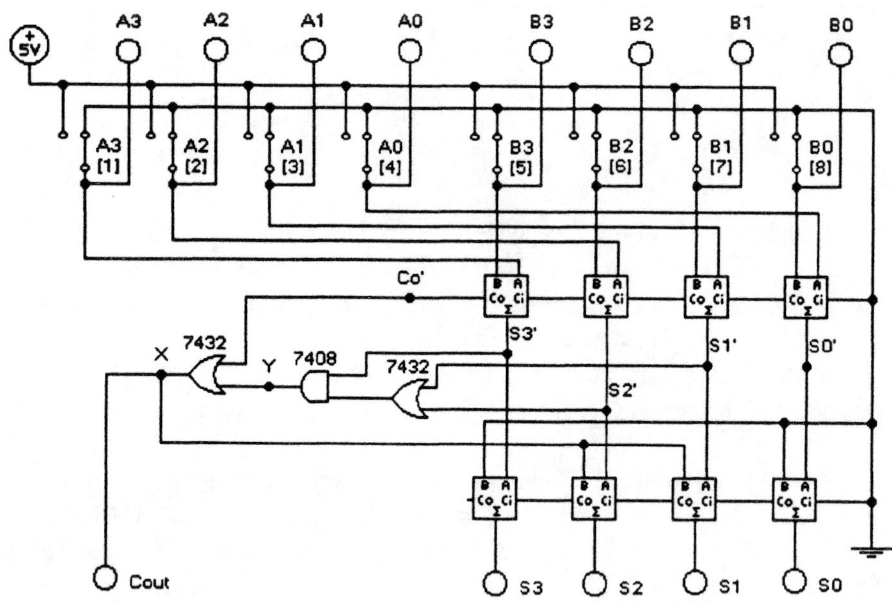

Figure 14-2 7483 BCD Adder

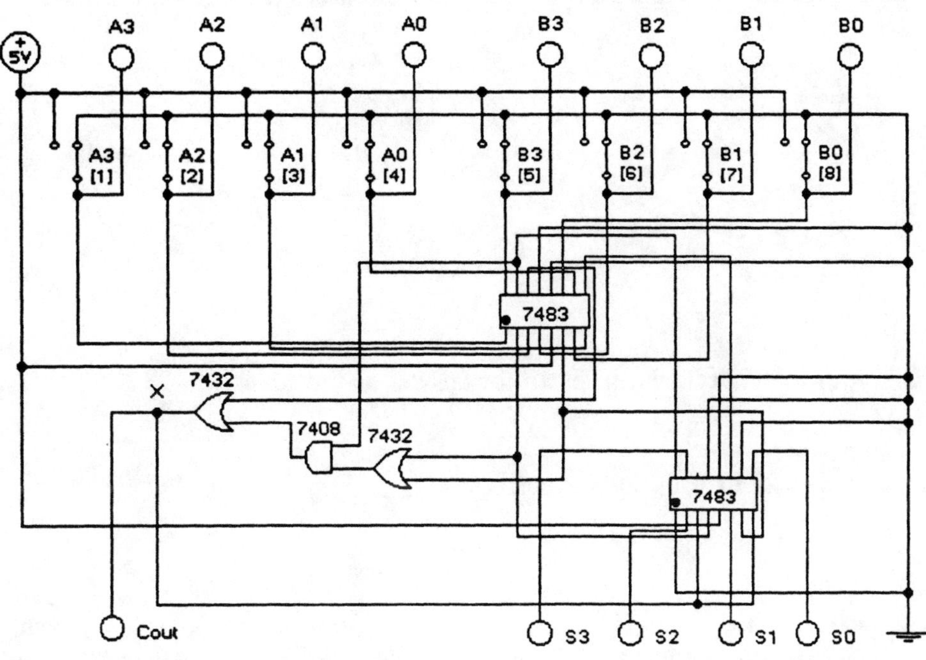

Procedure:

1. Pull down the File and open FIG14-1. Figure 14-1 shows a BCD adder constructed from two banks of four full-adders. See the Preparation section for more details. You will use this BCD adder to confirm the results when two BCD numbers are added. Press number keys 1-8 to control the logic switches. If you are performing this experiment in a hardwired laboratory, wire two 7483 four-bit binary adders as shown in Figure 14-2.

2. Add BCD 0111 to BCD 0010 in the space provided and confirm your results using the BCD adder in Figure 14-1 or Figure 14-2. Also show the decimal equivalent of the numbers and your answer.

Questions: Did your calculated answer match the answer on the adder?

Was the decimal equivalent of your answer correct?

3. Add BCD 1001 to BCD 0110 in the space provided and confirm your results using the BCD adder in Figure 14-1 or Figure 14-2. Also show the decimal equivalent of the numbers and your answer.

Questions: Did your calculated answer match the answer on the adder?

Was the decimal equivalent of your answer correct?

4. Add BCD 1001 to BCD 1000 in the space provided and confirm your results using the BCD adder in Figure 14-1 or Figure 14-2. Also show the decimal equivalent of the numbers and your answer.

Questions: Did your calculated answer match the answer on the adder?

Was the decimal equivalent of your answer correct?

5. Convert decimal 7 and decimal 5 to BCD numbers and add them in the space provided. Confirm your results using the BCD adder in Figure 14-1 or Figure 14-2.

Questions: Did your calculated answer match the answer on the adder?

Was the decimal equivalent of your answer correct?

6. Use a 4-input truth table with inputs S3', S2', S1', and S0' to prove that the AND-OR logic network in Figure 14-1 will produce a binary one (1) at the output of the 7408 AND gate (Y) when the binary output of the first adder is greater than nine (1001).

Name ______________________________
Date ______________________________

Parity Generator/Checker

Objectives:

1. Demonstrate how XOR gates are used to build a parity checker circuit.
2. Demonstrate how XOR gates are used to build a parity generator circuit.
3. Demonstrate how the 74280 parity generator/checker is used to build a parity error-detection system.
4. Demonstrate the difference between an even and an odd parity system.

Materials:

One 5 V dc voltage supply
Five logic switches
Eleven logic probe lights
One INVERTER (1-7404 IC)
Three XOR gates (1-7486 IC)
Two parity generator/checkers (2-74280 ICs)

Preparation:

When digital codes are being transmitted from one point to another, line noise can introduce an error in one of the bits. This may cause a binary one to be received as a binary zero or a binary zero to be received as a binary one. One of the methods used to detect this type of error is the parity method of error detection. In this method of error detection, a parity bit is added to the transmitted binary code to make the total number of binary ones even or odd (depending on the system). If the number of binary ones received is not the same parity (even or odd) as the parity of the transmitted code, there is a high probability that a transmission error has occurred.

The type of parity system (even or odd) must be the same at both the transmitter and the receiver so that the parity checker at the receiver knows whether to check for even or odd parity. The parity method of error detection is not perfect, but it has a high probability of success. More reliable methods of error detection are available but they are more costly to implement.

If the system is an even parity system, the parity generator will generate a binary one parity bit when the transmitted code has an odd number of binary ones. This will make the transmitted bits have an even number of binary ones. The parity generator will generate a binary zero parity bit when the transmitted code has an even number of binary ones, keeping the number of transmitted binary ones even. If the system is an odd parity system, the parity generator will generate a binary one parity bit when the transmitted code is even parity, and a binary zero parity bit will be generated when the transmitted code is odd parity.

A parity checker can be constructed using XOR gates because an XOR gate produces a binary one output when there is an odd number of binary ones on the input terminals. A 4-bit parity checker using three XOR gates is shown in Figure 15-1. The XOR gate output is used to detect an odd number of binary ones at the input. The inverted XOR gate output is used to detect an even number of binary ones at the input.

A parity generator can be constructed using XOR gates, as shown in Figure 15-2. When there is an odd number of binary ones input to the parity generator, the parity generator circuit (XOR gates and INVERTER) will produce a binary zero parity bit. This will cause an odd number of binary ones to be transmitted (odd parity). When there is an even number of binary ones input to the parity generator, the parity generator circuit will produce a binary one parity bit, causing an odd number of binary ones to be transmitted again (odd parity). This means that the parity generator circuit in Figure 15-2 is an odd parity generator.

The 74280 parity generator/checker, shown in Figure 15-3, can be used as a parity generator or a parity checker. When there is an odd number of binary ones on the 74280 inputs (A–I), the odd output will go to a binary one. When there is an even number of binary ones on the 74280 inputs (A–I), the even output will go to a binary one. When the 74280 is used as a parity generator, the even or odd output is used to generate the parity bit, depending on whether the system is even or odd parity. For an even parity system, the odd parity output is used. This will make the total number of binary ones transmitted always be even. For an odd parity system, the even parity output is used. This will make the total number of binary ones transmitted always be odd.

Figure 15-1 Four-Bit Parity Checker

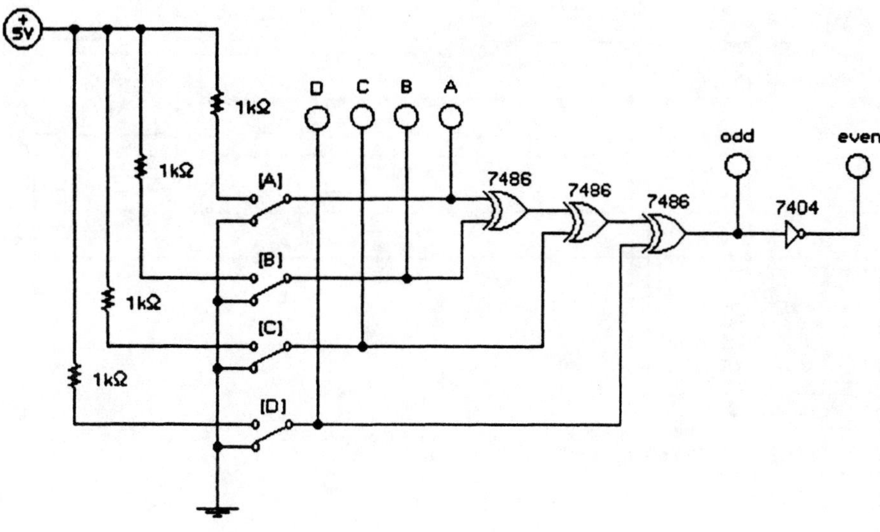

Figure 15-2 Four-Bit Parity Generator

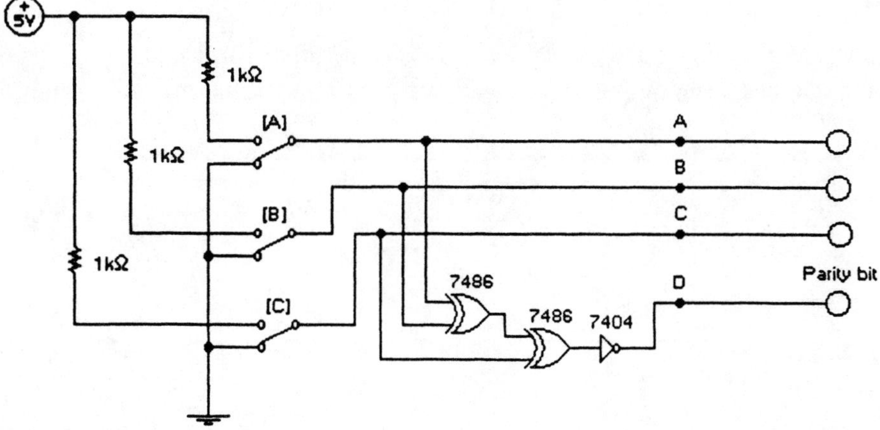

Figure 15-3 Parity Error-Detection System

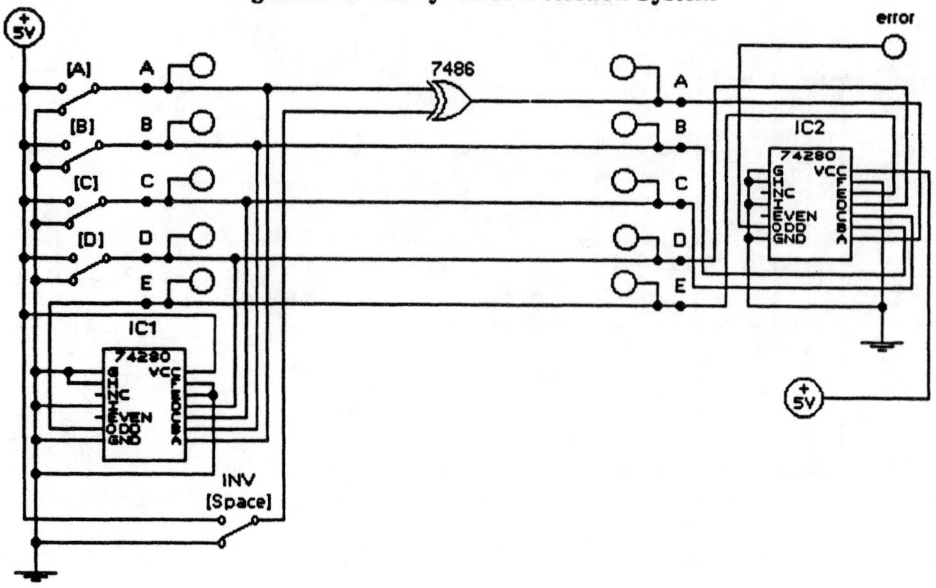

Procedure:

1. Pull down the File menu and open FIG15-1. For the circuit in Figure 15-1, record the predicted odd and even output logic levels for each input combination in Table 15-1.

Table 15-1

				Predicted		Measured	
D	C	B	A	odd	even	odd	even
0	0	0	0				
0	0	0	1				
0	0	1	0				
0	0	1	1				
0	1	0	0				
0	1	0	1				
0	1	1	0				
0	1	1	1				
1	0	0	0				
1	0	0	1				
1	0	1	0				
1	0	1	1				
1	1	0	0				
1	1	0	1				
1	1	1	0				
1	1	1	1				

2. Click the On-Off switch to run the analysis. By switching the logic switches to the appropriate positions, record the measured odd and even output logic levels for each input combination in Table 15-1. You can switch the logic switches by pressing the letter on the computer keyboard corresponding to the letter label on the switch.

Questions: How did your predicted outputs compare with the measured values?

What is the logic circuit in Figure 15-1 doing?

3. Click the On-Off switch to stop the analysis run. Pull down the File menu and open FIG15-2. For the circuit in Figure 15-2, record the predicted parity bit output logic levels for each input combination in Table 15-2.

Table 15-2

C	B	A	Predicted Parity Bit	Measured Parity Bit
0	0	0		
0	0	1		
0	1	0		
0	1	1		
1	0	0		
1	0	1		
1	1	0		
1	1	1		

4. Click the On-Off switch to run the analysis. By switching the logic switches to the appropriate positions, record the measured parity bit output logic levels for each input combination in Table 15-2. You can switch the logic switches by pressing the letter on the computer keyboard corresponding to the letter label on the switch.

Questions: How did your predicted outputs compare with the measured values?

What is the logic circuit in Figure 15-2 doing?

5. Click the On-Off switch to stop the analysis run. Pull down the File menu and open FIG15-3. Notice that a 4-bit binary code is generated by the settings of switches A, B, C, and D, with a fifth parity bit (bit E) generated by a parity generator (IC1). This 5-bit code is being transmitted over a 5-bit transmission line to a 5-bit parity checker (IC2) on the other end of the line. Also notice that a controlled inverter (7486) has been inserted in one of the transmission lines (line A) to make it possible to simulate a bit error. When the INV switch is down (0), the binary bit on line A is passed through the controlled inverter unchanged. When the INV switch is up (1), the binary bit on line A is inverted, causing the output code to have a different parity than the original transmitted code. This simulates an error in the

data transmission, causing the error light on the parity checker to light. Review the Preparation section of this experiment for further discussion of the 74280 parity generator/checker.

6. Click the On-Off switch to run the analysis. By pressing the A, B, C, and D keys on the computer keyboard, test each 4-bit input combination to generate the various binary codes.

Questions: Were the received binary codes the same as the transmitted binary codes?

Did the error light turn on for any of the transmitted codes? Explain the results.

7. Press the space bar on the keyboard to switch the INV switch to the up (1) position. By pressing the A, B, C, and D keys on the computer keyboard, test each 4-bit input combination to generate the various binary codes again.

Questions: Were the received binary codes the same as the transmitted binary codes?

Did the error light turn on for any of the transmitted codes? Explain the results.

Is this an even or an odd parity error-detection system? Explain your answer.

Name ______________________

Date ______________________

Magnitude Comparator

Objectives:

1. Demonstrate how the XNOR gate (XOR gate and INVERTER) is used to build a magnitude comparator that detects two equal 2-bit binary numbers.
2. Demonstrate how the XNOR gate is used to build a magnitude comparator that detects two equal 4-bit binary numbers.

Materials:

One 5 V dc voltage supply
Eight logic switches
Nine logic probe lights
Four INVERTERS (1-7404 IC)
Four XOR gates (1-7486 IC)
One two-input AND gate (1-7408 IC)
One four-input AND gate (1-7421 IC)

Preparation:

The function of a digital magnitude comparator is to compare the magnitudes of two binary numbers to determine their relationship. In its basic form, a magnitude comparator determines whether two binary numbers are equal. Most integrated circuit magnitude comparators, such as the 7485, have three outputs. One output produces a binary one when the binary numbers are equal, the second output produces a binary one when the A binary number is greater than the B number, and the third output produces a binary one when the B binary number is greater than the A number. In this experiment, you will demonstrate how the XNOR gate can be used to produce a binary one output when two binary numbers are equal in magnitude. Review the discussion of the XNOR gate in the Preparation section of Experiment 11 before proceeding with this experiment.

Because the XNOR gate (XOR with an INVERTER on the output) produces a binary one output when both inputs are equal, it is an ideal gate for building a magnitude comparator. The circuit in Figure 16-1 will be used to demonstrate a 2-bit magnitude comparator using XNOR gates. Notice that the two least-significant bits of the 2-bit binary numbers are applied to the inputs of one XNOR gate and the two most-significant bits of the 2-bit binary numbers are applied to the inputs of the other XNOR gate. When the two binary numbers are equal, their corresponding bits will be equal, causing both XNOR gates to produce a binary one output to both AND gate inputs. This will cause the AND gate to produce a binary one output, lighting the logic probe light (A = B).

The circuit in Figure 16-2 will be used to demonstrate a 4-bit magnitude comparator using XNOR gates. This circuit will compare two 4-bit binary numbers and produce a binary one output when both 4-bit numbers are equal. The theory for this 4-bit magnitude comparator is the same as the theory for the 2-bit magnitude comparator, described previously. Notice that A0 and B0 are applied to the first XNOR gate, A1 and B1 are applied to the second XNOR gate, A2 and B2 are applied to the third XNOR gate, and A3 and B3 are applied to the fourth XNOR gate.

Figure 16-1 Two-Bit Magnitude Comparator

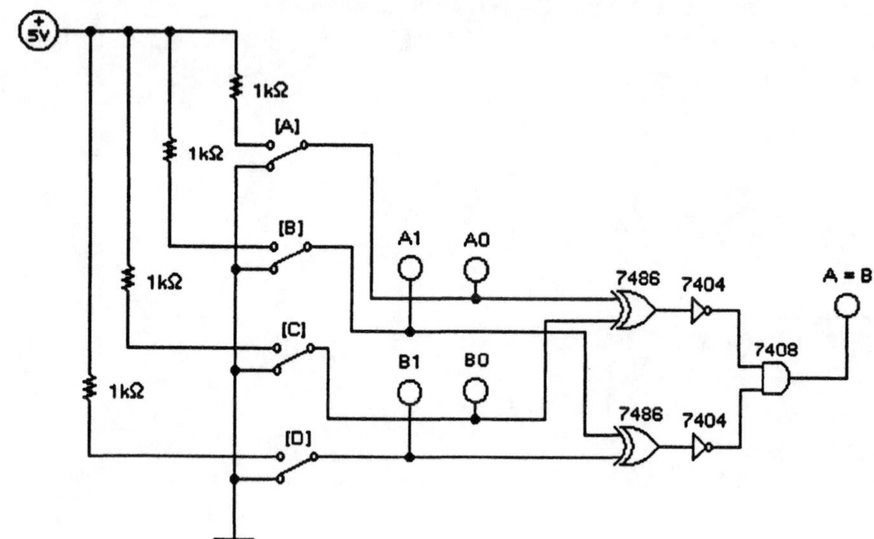

Figure 16-2 Four-Bit Magnitude Comparator

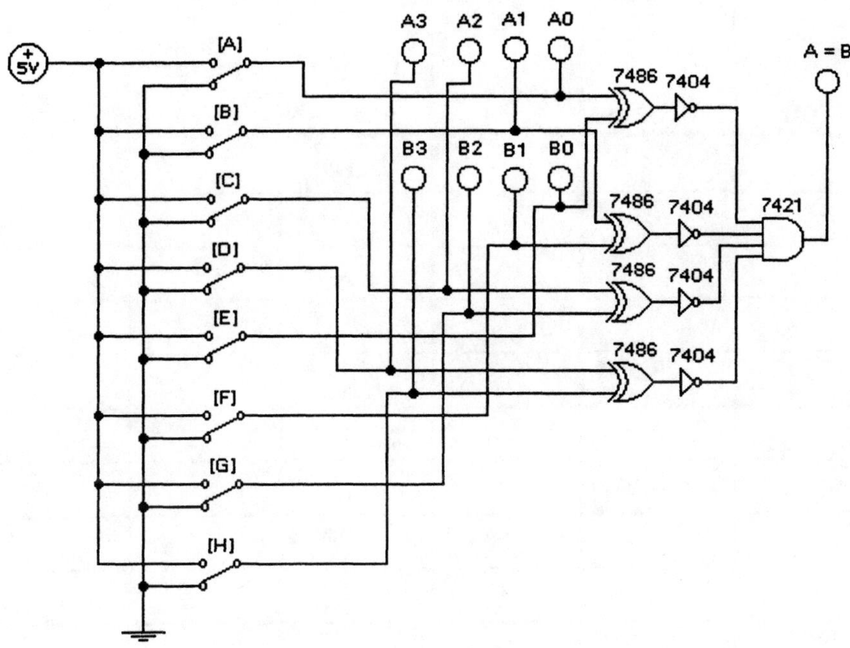

Procedure:

1. Pull down the File menu and open FIG16-1. For the circuit in Figure 16-1, record the predicted output (A = B) for each input combination in Table 16-1.

Table 16-1

				Predicted	Measured
B1	B0	A1	A0	A=B	A=B
0	0	0	0		
0	0	0	1		
0	0	1	0		
0	0	1	1		
0	1	0	0		
0	1	0	1		
0	1	1	0		
0	1	1	1		
1	0	0	0		
1	0	0	1		
1	0	1	0		
1	0	1	1		
1	1	0	0		
1	1	0	1		
1	1	1	0		
1	1	1	1		

2. Click the On-Off switch to run the analysis. By switching the logic switches to the appropriate positions, record the measured output (A = B) logic levels for each input combination in Table 16-1. You can switch the logic switches by pressing the letter on the computer keyboard corresponding to the letter label on the switch.

Questions: How did your predicted outputs compare with the measured values?

What is the logic circuit in Figure 16-1 doing?

3. Click the On-Off switch to stop the analysis run. Pull down the File menu and open FIG16-2. For the circuit in Figure 16-2, record the predicted output (A = B) for each input combination in Table 16-2.

Table 16-2

B3	B2	B1	B0	A3	A2	A1	A0	Predicted A=B	Measured A=B
0	0	0	0	0	0	0	0		
0	0	0	1	0	0	0	1		
0	0	1	0	0	0	1	0		
0	0	1	1	0	0	1	1		
0	1	0	0	0	1	0	0		
0	1	0	1	0	1	0	1		
0	1	1	0	0	1	1	0		
0	1	1	1	0	1	1	1		
1	0	0	0	0	0	0	0		
1	0	0	1	0	0	0	1		
1	0	1	0	0	0	1	0		
1	0	1	1	0	0	1	1		
1	1	0	0	0	1	0	0		
1	1	0	1	0	1	0	1		
1	1	1	0	0	1	1	0		
1	1	1	1	0	1	1	1		

4. Click the On-Off switch to run the analysis. By switching the logic switches to the appropriate positions, record the measured output (A = B) logic levels for each input combination in Table 16-2. You can switch the logic switches by pressing the letter on the computer keyboard corresponding to the letter label on the switch.

Questions: How did your predicted outputs compare with the measured values?

What is the logic circuit in Figure 16-2 doing?

Name ______________________________

Date ______________________________

EXPERIMENT

17 Troubleshooting Arithmetic Circuits

Objectives:

1. Determine the defective logic gate or component for various arithmetic logic circuits by monitoring the logic levels at circuit test points.

Materials:

This experiment can only be performed on Electronics Workbench using the circuits disk provided with this manual.

Preparation:

In order to perform this experiment effectively, you must first complete Experiments 11–16. Use the theory learned in those experiments to find the defective logic gate or component in the arithmetic circuits in this experiment.

Determine the defective component in each experiment by changing the logic circuit binary inputs until you find the binary inputs that cause the circuit output to be incorrect. Observe the inputs and outputs of each logic gate when the circuit output is incorrect. A logic gate is defective if it has an incorrect output for any combination of binary inputs, unless one of the inputs is connected to a defective logic gate with an open output. Remember that an open input on a 7400 series logic gate will behave as if there is a logical "one" on that open input terminal, even if that open input is being caused by the open output of another logic gate and the logic probe light is indicating a zero. For example, if one of the inputs of an OR gate that is not defective is connected to a logic gate with an open output, the OR gate will act as if there is a logical "one" on that input and a logical "one" on its output (logic probe light "on"), even if all of the OR gate inputs indicate logical "zero" (logic probe light "off"). This will happen because an OR gate produces a logical "one" at the output when any input is at a logical "one" (or open). For this reason, you may need to try all possible logic network binary input combinations before concluding which logic gate is defective.

Procedure:

Don't forget to read the Preparation section before attempting to determine the defective components in the following arithmetic circuits.

1. Pull down the File menu and open FIG17-1. Click the On-Off switch to run the analysis. Based on the logic levels at the full-adder test points, determine which logic gate is defective. To switch the logic switches, press the key on the computer keyboard that matches the letter label on the switch.

   Defective gate ________________

2. Pull down the File menu and open FIG17-2. Click the On-Off switch to run the analysis. Based on the logic levels at the full-adder test points, determine which logic gate is defective. To switch the logic switches, press the key on the computer keyboard that matches the letter label on the switch.

   Defective gate ________________

3. Pull down the File menu and open FIG17-3. Click the On-Off switch to run the analysis. Based on the logic levels at the full-adder test points, determine which logic gate is defective. To switch the logic switches, press the key on the computer keyboard that matches the letter label on the switch.

   Defective gate ________________

4. Pull down the File menu and open FIG17-4. Click the On-Off switch to run the analysis. Based on the logic levels at the full-adder test points, determine which logic gate is defective. To switch the logic switches, press the key on the computer keyboard that matches the letter label on the switch.

   Defective gate ________________

5. Pull down the File menu and open FIG17-5. Click the On-Off switch to run the analysis. Based on the logic levels at the full-adder test points, determine which logic gate is defective. To switch the logic switches, press the key on the computer keyboard that matches the letter label on the switch.

   Defective gate ________________

6. Pull down the File menu and open FIG17-6. Click the On-Off switch to run the analysis. Using the logic probe light to measure the logic levels at various test points in the full-adder circuit, determine which logic gate is defective. To switch the logic switches, press the key on the computer keyboard that matches the letter label on the switch.

   Defective gate ______________

7. Pull down the File menu and open FIG17-7. Click the On-Off switch to run the analysis. Using the logic probe light to measure the logic levels at various test points in the full-adder circuit, determine which logic gate is defective. To switch the logic switches, press the key on the computer keyboard that matches the letter label on the switch.

   Defective gate ________________

8. Pull down the File menu and open FIG17-8. Click the On-Off switch to run the analysis. Using the logic probe light to measure the logic levels at various test points in the full-adder circuit, determine which logic gate is defective. To switch the logic switches, press the key on the computer keyboard that matches the letter label on the switch.

   Defective gate ________________

9. Pull down the File menu and open FIG17-9. Click the On-Off switch to run the analysis. Using the logic probe light to measure the logic levels at various test points in the magnitude comparator circuit, determine which logic gate is defective. To switch the logic switches, press the key on the computer keyboard that matches the letter label on the switch.

   Defective gate ________________

10. Pull down the File menu and open FIG17-10. Click the On-Off switch to run the analysis. Using the logic probe light to measure the logic levels at various test points in the adder/subtractor circuit, determine which component is defective. To switch the logic switches, press the key on the computer keyboard that matches the number label on the switch.

    Defective component ________________

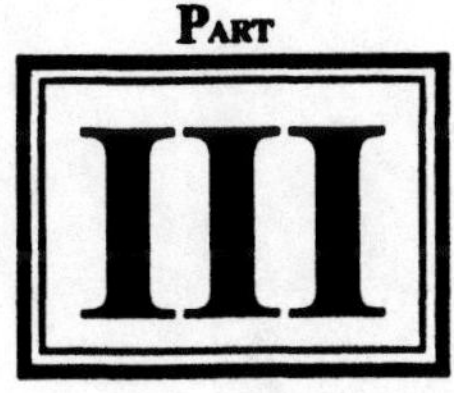

MSI Logic Circuits

The experiments in Part III involve the study of medium scale integrated circuits such as decoders, encoders, multiplexers, and demultiplexers. In the final experiment in Part III you will solve some troubleshooting problems involving these devices.

The circuits for the experiments in Part III can be found on the enclosed disk in the PART3 subdirectory.

Name ____________________
Date ____________________

Decoders and Encoders

Objectives:

1. Investigate the operation of a 74138 3 line-to-8 line decoder.
2. Demonstrate how to wire a 4 line-to-16 line decoder using two 74138 3 line-to-8 line decoders.
3. Investigate the operation of a 7442 BCD-to-decimal decoder.
4. Investigate the operation of a BCD-to-7 segment decoder/driver driving a 7 segment LED display.
5. Investigate the operation of a 74147 decimal-to-BCD priority encoder.
6. Demonstrate an encoder-decoder logic circuit using a 74148 3 line-to-8 line encoder and a BCD-to-7 segment decoder/driver.
7. Design a 2 line-to-4 line decoder using AND gates and inverters.

Materials:

One 5 V dc power supply
Ten logic switches
Twenty logic probe lights
Two 3 line-to-8 line decoders (2-74138 ICs)
One BCD-to-decimal decoder (1-7442 IC)
One BCD-to-7 segment decoder/driver (1-7448 IC)
One common cathode 7 segment LED display
One decimal-to-BCD encoder (1-74147 IC)
Four INVERTERS (1-7404 IC)
One 3 line-to-8 line encoder (1-74148 IC)
Four two-input AND gates (1-7408 IC)

Preparation:

A decoder is a logic circuit that accepts a binary input and activates only the output that corresponds to the binary input. Some decoders have one or more enable inputs that are used to enable or disable the decoder.

The 74138 in Figure 18-1 is a 3 line-to-8 line octal decoder capable of decoding eight octal codes (000–111) into eight separate active low outputs. It also has three enable inputs (G1, G2A, and

G2B). Enable input G1 must be high (1) and enable inputs G2A and G2B must be low (0) to enable the 74138 decoder. When the decoder is disabled, all outputs are at a logical high (1). When the decoder is enabled, the binary code on inputs A, B, and C determines which output will go low (0).

The circuit in Figure 18-2 will demonstrate the timing of the 74138 decoder output waveforms using a logic analyzer and a word generator. The word generator will transmit binary numbers from 000 to 111 in a two cycle sequence while the logic analyzer monitors the eight output terminals. If this part of the experiment is performed in a hardwired laboratory, a binary counter can be used in place of the word generator.

The circuit in Figure 18-3 shows how to connect two 74138 3 line-to-8 line decoders to make a 4 line-to-16 line (1-of-16) decoder. The fourth input (D) is connected to enable terminal G1 on the second 74138 decoder and enable terminal G2A (inverted input) on the first 74138 decoder. Therefore, the first decoder will display outputs for binary inputs 0000 through 0111 when D is binary zero (0), and the second decoder will display outputs for binary inputs 1000 through 1111 when D is binary one (1).

The 7442 in Figure 18-4 is a BCD-to-decimal (4 line-to-10 line) decoder with 10 active low outputs (0–9). When an input (DCBA) outside the range of zero (0000) to nine (1001) is applied, all of the outputs remain high (1) because the BCD code only includes binary numbers between zero and nine. The 7442 does not have an enable input. Therefore, one of the outputs will go low (0) as soon as a binary input between zero and nine is applied.

The circuit in Figure 18-5 is a BCD-to-7 segment decoder/driver connected to a 7 segment LED display. The LEDs will display the decimal equivalent of the BCD code applied to the decoder/driver input (DCBA) for numbers between zero and nine. Numbers outside the range of zero to nine do not exist in the BCD code. Because the decoder/driver has active high outputs, a common cathode LED display is required. External pull-up resistors are not needed because they are built into the decoder/driver circuit. If this part of the experiment is performed in a hardwired laboratory, a 7448 decoder/driver should be used with a common cathode 7 segment LED display. Consult the 7448 and LED display literature for wiring instructions and pin numbers.

Encoding is the opposite process from decoding. An encoder generates a binary coded output from a singular active input. Some encoders have an enable input that may be used to enable or disable the encoder. Some encoders also have an output that is active only when none of the inputs are active. This output can be used to control external logic circuitry or blank an LED display.

The 74147 in Figure 18-6 is a decimal-to-BCD priority encoder with active low inputs and active low outputs. A priority encoder will only respond to the input that represents the value with the highest priority (usually the highest value) when more than one input is active. Because of the active low outputs, INVERTERS are required for each output terminal. Without the INVERTERS, the encoder would produce an inverted BCD output code. The 74147 has only nine inputs (input zero is missing). Input zero (0) is not connected to the 74147 encoder because the encoder assumes a decimal zero (0) input and will output a BCD zero (0000) code when none of the inputs are active.

The circuit in Figure 18-7 will demonstrate how an encoder and a 7 segment decoder/driver and LED display can be used to display the decimal equivalent of a key closure. A 74148 8 line-to-3 line priority encoder is used because it has an output (E0) that will go high (1) when one of the inputs is active and stays low (0) when none of the inputs are active. This output (E0) is used to blank the 7 segment display until one of the encoder inputs is active. This is accomplished by connecting the encoder E0 output to the decoder/driver active low blanking input (BI). When none of the encoder inputs are active, the encoder E0 output will bring the decoder/driver active low blanking input (BI) to binary zero (0), blanking the 7 segment display. When one of the encoder inputs is active, encoder output E0 will bring the decoder/driver blanking input (BI) to binary one (1), allowing the decimal output to be displayed on the 7 segment LED. Because the 74148 encoder has only three outputs, input D of the decoder/driver is connected to ground (0). Therefore, only decimal numbers 0–7 can be displayed. The 74148 encoder also has an active low enable input (Ei). Therefore, this input (Ei) must be connected to ground (0) to enable the 74148. Similar to the 74147, the 74148 has active low inputs and outputs. Therefore, INVERTERS are required on the 74148 encoder outputs also. When more than one of the inputs are active (low), the 74148 encoder output is controlled by the input that represents the highest decimal value because the 74148 is a priority encoder.

Figure 18-1 3 Line-to-8 Line (1-of-8) Decoder

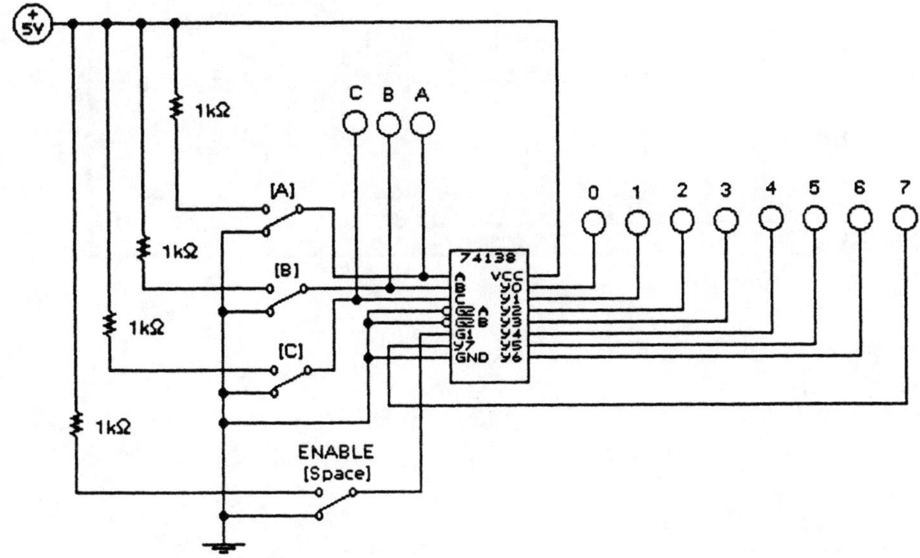

Figure 18-2a Decoder Output Waveform Analysis (EWB Version 4)

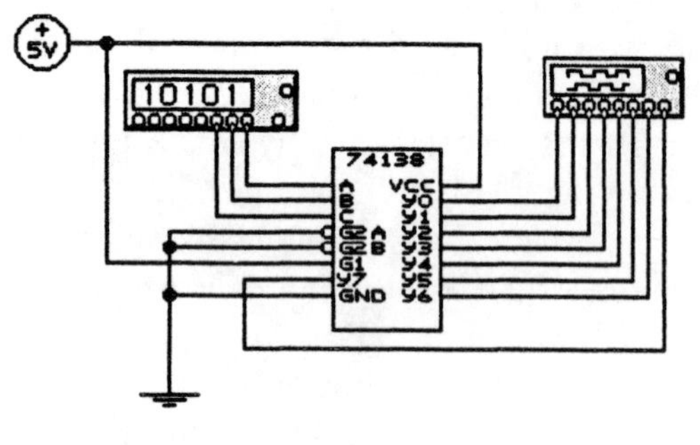

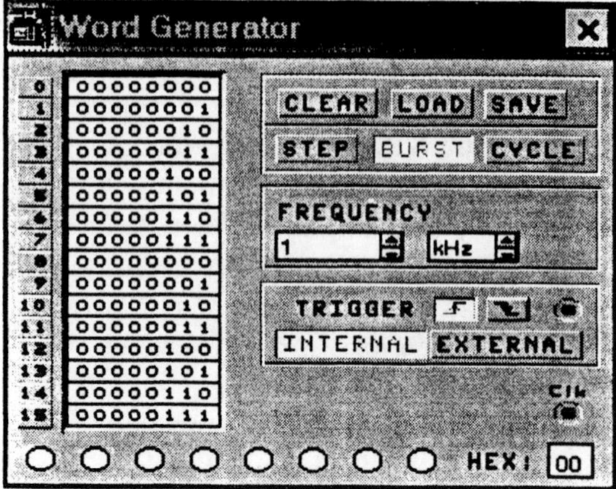

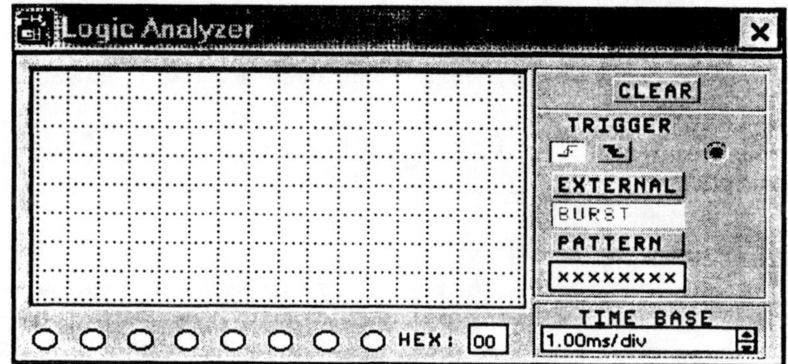

Figure 18-2b Decoder Output Waveform Analysis (EWB Version 5)

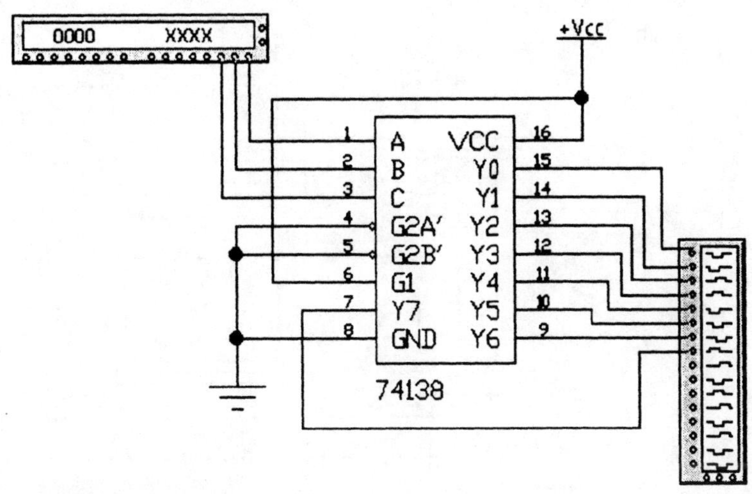

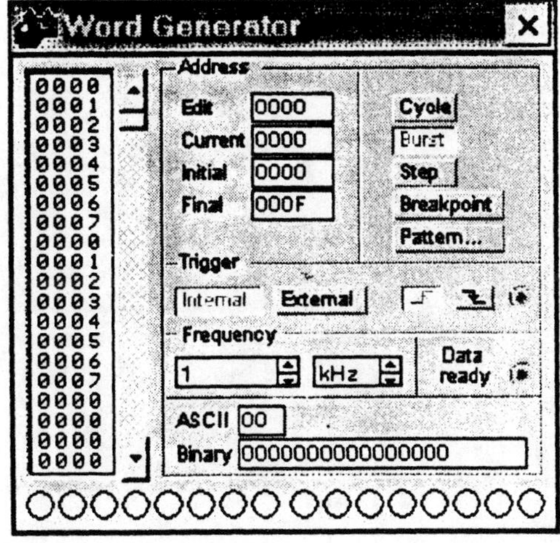

Logic Analyzer Settings
Clocks per division ---------------- 8

Clock Setup dialog box
Clock edge --------------- positive
Clock mode -------------- Internal
Internal clock rate ------ 10 kHz
Clock qualifier ----------- x
Pre-trigger samples --- 100
Post-trigger samples -- 1000
Threshold voltage (V) 2

Trigger Patterns dialog box
A ------------------------------ xxxxxxxxxxxxxxxx
Trigger combinations -- A
Trigger qualifier ---------- x

Figure 18-3 4 Line-to-16 Line (1-of-16) Decoder

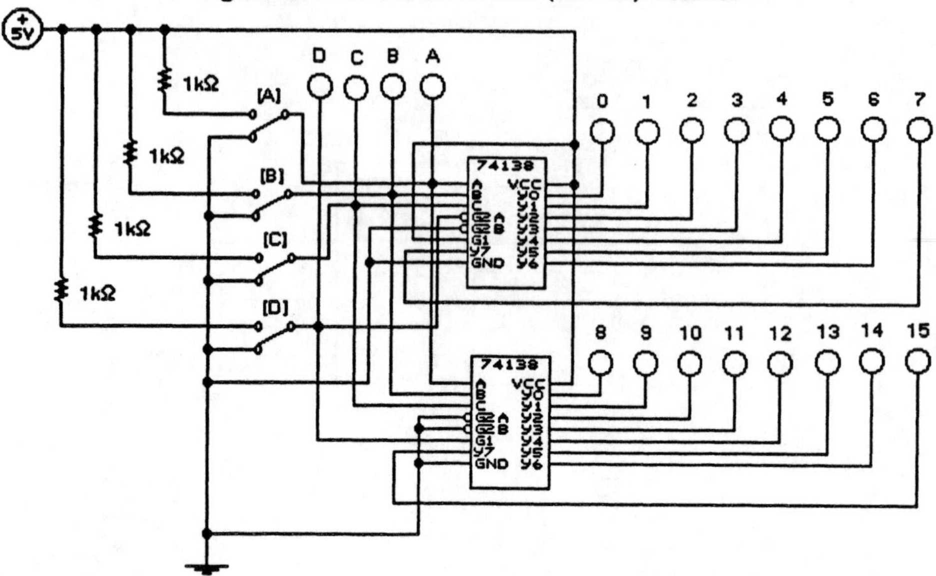

Figure 18-4 BCD-to-Decimal Decoder

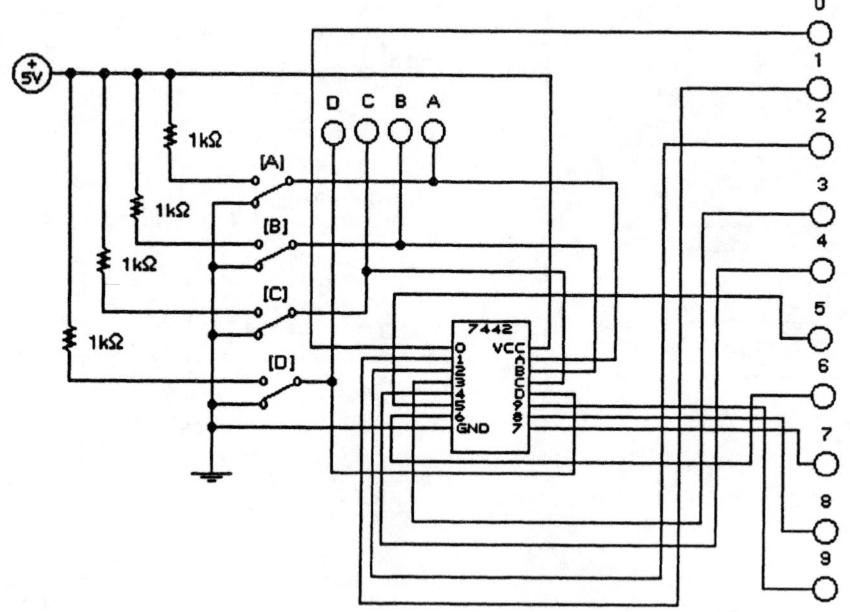

Figure 18-5 BCD-to-7 Segment Decoder/Driver

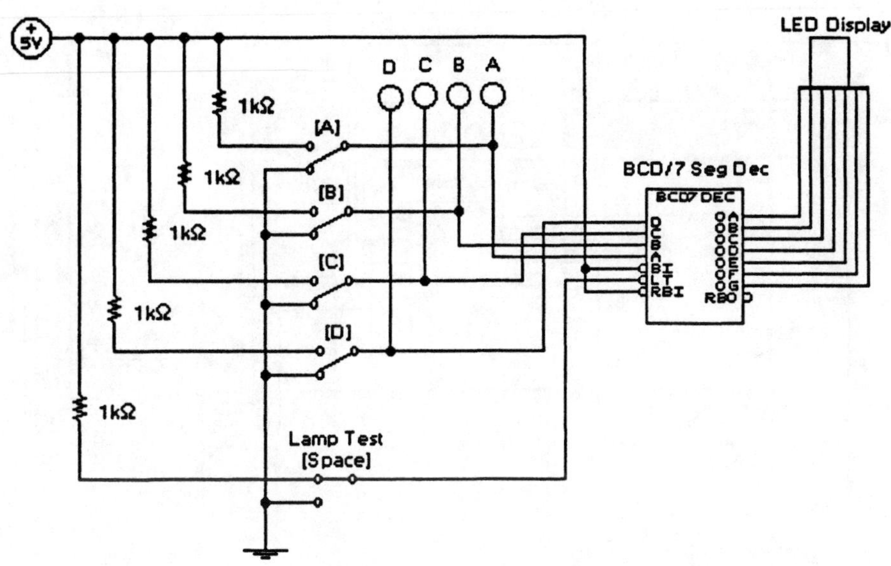

Figure 18-6 Decimal-to-BCD Priority Encoder

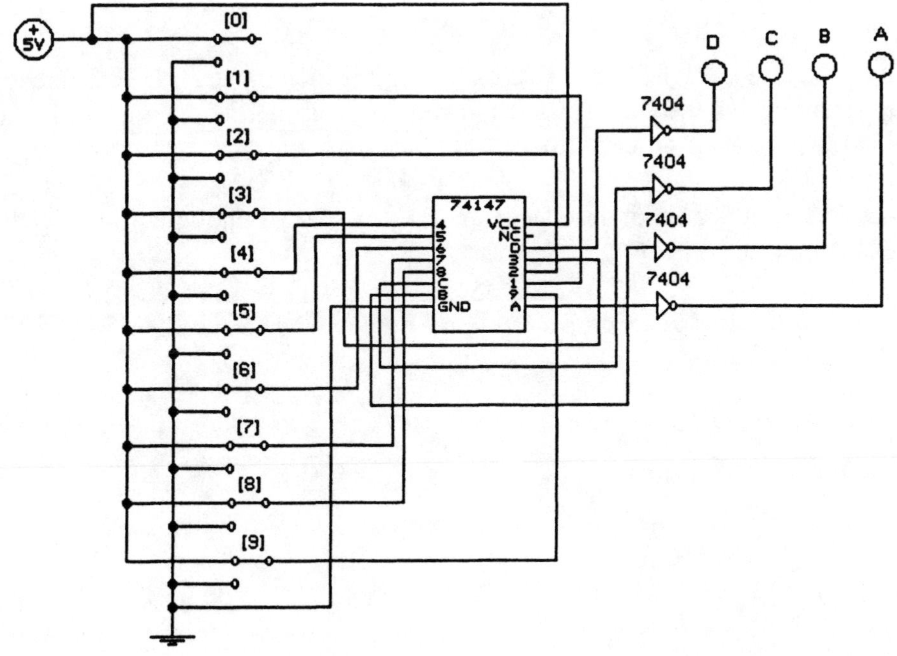

Figure 18-7 Encoder/Decoder Circuit

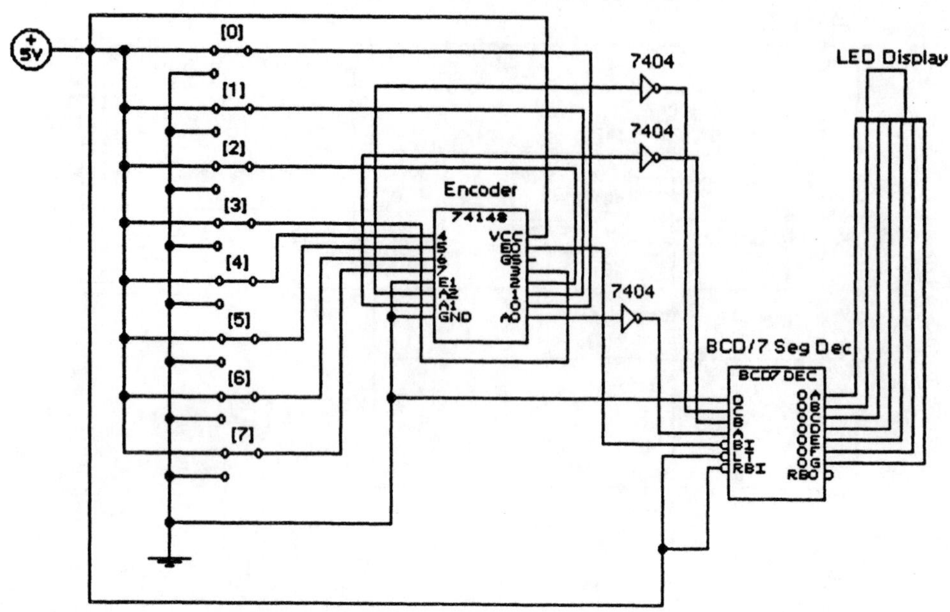

Procedure:

1. Pull down the File menu and open FIG18-1. You are looking at a 74138 3 line-to-8 line decoder circuit. Click the On-Off switch to run the analysis. Notice that enable terminal G1 on the 74138 decoder is grounded (0).

Question: What do you notice about decoder outputs Y0–Y7 based on logic probe lights 0–7? Is the decoder enabled or disabled?

2. Set the ENABLE switch to binary one (up) by pressing the space bar on the computer keyboard. This places a binary one (1) on the 74138 decoder enable terminal G1. Notice that enable terminals G2A and G2B on the decoder are grounded (0). Also notice that the binary input (CBA) is 000.

Question: What do you notice about decoder outputs Y0–Y7 based on logic probe lights 0–7? Explain your answer.

3. By pressing the A, B, and C keys on the keyboard, change the binary input to the values in Table 18-1 and record the outputs in the table.

Table 18-1 74138 Decoder

Inputs			Outputs							
C	B	A	Y0	Y1	Y2	Y3	Y4	Y5	Y6	Y7
0	0	0								
0	0	1								
0	1	0								
0	1	1								
1	0	0								
1	0	1								
1	1	0								
1	1	1								

Question: What conclusion can you draw about the relationship between the decoder outputs and the binary inputs? Are the outputs active low or active high?

4. Click the On-Off switch to stop the analysis run. Pull down the File menu and open FIG18-2. The word generator and logic analyzer settings should be as shown in Figure 18-2. Notice that the decoder enable terminal G1 is connected to 5 V (1) and enable terminals G2A and G2B are grounded (0) to enable the decoder. Click BURST on the word generator to run the analysis. Notice that the decoder outputs Y0–Y7 are being monitored by the logic analyzer. (The top curve plot is Y0 and the bottom curve plot is Y7). Also notice that the

word generator transmitted binary 000 through binary 111 to the decoder input (CBA) in a two cycle sequence in the burst mode.

> NOTE: If this part of the experiment is performed in a hardwired laboratory, a binary counter can be used in place of the word generator.

Question: What is the relationship between the outputs (Y0–Y7) on the logic analyzer screen and the binary inputs generated by the word generator?

5. Click the On-Off switch to stop the analysis run. Pull down the File menu and open FIG18-3. You are looking at two 74138 3 line-to-8 line decoders wired as a 4 line-to-16 line decoder. Notice that the fourth input (input D) is connected to enable G1 on the bottom 74138 decoder and enable G2A (inverted input) on the top 74138 decoder. Also notice that inputs A, B, and C are connected to terminals A, B, and C on both 74138 decoders. Click the On-Off switch to run the analysis.

Question: What do the outputs indicate? Explain.

6. By pressing the A, B, C, and D keys on the keyboard, change the binary inputs and notice the outputs.

Question: What conclusion can you draw about the relationship between the decoder outputs and the binary inputs? Explain.

7. Click the On-Off switch to stop the analysis run. Pull down the File menu and open FIG18-4. You are looking at a 7442 BCD-to- decimal (4 line-to-10 line) decoder circuit. Notice that the 7442 does not have an enable input for enabling or disabling the decoder. Click the On-Off switch to run the analysis. By pressing the A, B, C, and D keys on the keyboard, change the binary input to the values in Table 18-2 and record the outputs in the table.

Table 18-2 7442 Decoder

Inputs				Outputs									
D	C	B	A	0	1	2	3	4	5	6	7	8	9
0	0	0	0										
0	0	0	1										
0	0	1	0										
0	0	1	1										
0	1	0	0										
0	1	0	1										
0	1	1	0										
0	1	1	1										
1	0	0	0										
1	0	0	1										
1	0	1	0										
1	0	1	1										
1	1	0	0										
1	1	0	1										
1	1	1	0										
1	1	1	1										

Questions: What conclusion can you draw about the relationship between the decoder outputs and the binary inputs? Are the outputs active low or active high?

What happens when a binary input outside the range of 0–9 is applied?

8. Click the On-Off switch to stop the analysis run. Pull down the File menu and open FIG18-5. You are looking at a BCD-to-7 segment decoder/driver connected to a 7 segment LED display. Because the decoder/driver has active high outputs, the LED display is a common cathode display. This decoder circuit will display the decimal equivalent of the BCD code applied to the decoder/driver input. Click the On-Off switch to run the analysis. Press the space bar on the keyboard to switch the Lamp Test switch to ground (down). This will place a binary zero (0) on the active low lamp test (LT) terminal on the decoder/driver, which will cause all of the LED segments to light on the 7 segment display if the display is functioning properly.

> NOTE: If this part of the experiment is performed in a hardwired laboratory, a 7448 decoder/driver should be used with a common cathode 7 segment LED display. Consult the 7448 and LED display literature for wiring instructions and pin numbers.

Question: Did the lamp test verify that your 7 segment LED display is working properly?

9. Press the space bar to place a binary one on the decoder active low lamp test (LT) terminal and remove the lamp test. By pressing the A, B, C, and D keys on the keyboard, try different binary inputs to verify that the LED display shows the correct decimal equivalents.

Questions: Does your LED display show the correct decimal numbers for the binary inputs 0000 through 1001?

What happens when a binary number outside the range of zero (0000) through nine (1001) is applied to the decoder input?

10. Click the On-Off switch to stop the analysis run. Pull down the File menu and open FIG18-6. You are looking at a decimal-to-BCD priority encoder circuit. Click the On-Off switch to run the analysis. Notice what happens to the binary output (DCBA) when you open and close the numbered switches connected to the 74147 encoder inputs. You can open and close the numbered switches by pressing the equivalent number keys on the computer keyboard.

Questions: What is the relationship between the numbered switches and the binary output (DCBA)?

Are the inputs on the 74147 encoder active low or active high?

What happens when two input number switches are active at the same time? Which input controls the BCD output? Explain.

Why is the input zero (0) switch not connected to the encoder? What is the 74147 encoder BCD output when none of the inputs are active?

Are the outputs on the 74147 encoder active low or active high? What is the purpose of the four INVERTERS connected to the encoder outputs?

11. Click the On-Off switch to stop the analysis run. Pull down the File menu and open FIG18-7. You are looking at a 74148 8 line-to-3 line priority encoder feeding a BCD-to-7 segment decoder. Because the decoder has four inputs (DCBA) and the encoder has three outputs, input D of the decoder is connected to ground (0). Therefore, the decoder will only count to seven (binary 111). Also, output E0 of the 74148 encoder produces a binary zero (0) output when none of the encoder inputs are active. Because E0 is connected to the active low blanking input (BI) of the decoder, the decoder output will be blanked-out when none of the encoder inputs are active. Therefore, the LED display will only display a decimal number when one of the encoder input number keys is active (low). Click the On-Off switch to run the analysis. Activate the number switches by pressing the equivalent number keys on the keyboard and notice the readings on the LED display.

Questions: What is the relationship between the numbered switches and the decimal display?

Are the inputs on the 74148 encoder active low or active high?

What happens when two input number switches are active at the same time? Which input controls the decimal output? Explain.

What happens when none of the number switches are active (0)? Explain.

Are the outputs on the 74148 encoder active low or active high? What is the purpose of the three INVERTERS connected to the encoder outputs?

Is the enable (Ei) terminal on the 74148 encoder active low or active high?

12. Click the On-Off switch to stop the analysis run. Based on the truth table in Table 18-3, design a 2 line-to-4 line decoder with active high outputs using AND gates and INVERTERS. Design a separate logic circuit for each output, all connected to the same two inputs (B and A). Review the Preparation section of Experiment 9 before attempting this step. Show the logic network in the space provided.

Table 18-3 Decoder Design Problem

Inputs		Outputs			
B	A	Y0	Y1	Y2	Y3
0	0	1	0	0	0
0	1	0	1	0	0
1	0	0	0	1	0
1	1	0	0	0	1

13. Wire the logic network drawn in Step 12 using Electronics Workbench (or hardwired) and verify that it matches the truth table in Table 18-3. Use logic probe lights to monitor the outputs and logic switches for inputs A and B.

Question: Did your logic circuit satisfy the requirements of the truth table in Table 18-3? If not, what changes are required?

Name ______________________

Date ______________________

Multiplexers and Demultiplexers

Objectives:

1. Investigate the operation of an 8-input multiplexer.
2. Investigate the operation of a quad 2-input multiplexer.
3. Demonstrate how a multiplexer can be used to transmit serial data.
4. Demonstrate how a multiplexer can be used as a logic function generator to simulate various logic gates and logic networks.
5. Demonstrate how a decoder is used as a demultiplexer.
6. Demonstrate an application for a multiplexer and demultiplexer.

Materials:

One 5 V dc power supply
Eleven logic switches
Eleven logic probe lights
One pulse generator
One logic analyzer or oscilloscope
One eight-input multiplexer (1-74151 IC)
One quad two-input multiplexer (1-74157 IC)
One BCD-to-7 segment decoder/driver (1-7448 IC)
One common cathode 7 segment LED display
One decoder/demultiplexer (1-74138 IC)

Preparation:

A digital multiplexer (data selector) is a logic circuit that accepts a number of digital data inputs from several sources and connects one of them to a single output. The multiplexer inputs and output may be single lines each or multiple lines each. The selection of which digital input will be connected to the output is determined by the binary code applied to the SELECT input. Some multiplexers have one or more enable inputs that are used to enable or disable the multiplexer. A multiplexer acts like a digitally controlled multiposition switch.

The 74151 in Figure 19-1 is an 8-input multiplexer with single line inputs (D7–D0) and a single line output (Y). It also has an active low enable input (G′) and an inverted output (W). The active low enable input (G′) must be low (connected to ground) to enable the 74151 multiplexer. The SELECT inputs (CBA) require a binary number to select which input will be connected to the output. For example, if binary zero (000) is applied to the SELECT input (CBA), input D0 will connect to the output (Y).

The circuit in Figure 19-2 is a 74157 quad 2-input multiplexer wired to a BCD decoder/driver and a 7 segment LED display. The 74157 has two 4-line inputs (A and B) and a 4-line output (Y). The four multiplexer output lines are connected to the BCD decoder/driver input (DCBA). The 74157 multiplexer SELECT input (A/B′) is a single line because only one of two inputs needs to be selected. The 74157 has an active low enable (G′) that must be connected to ground (0) to enable the multiplexer. (Note: Make sure you complete Experiment 18 on decoders before continuing with this experiment).

The circuit in Figure 19-3 will demonstrate how a multiplexer is used as a parallel-to-serial converter that transmits a parallel binary word over a single line one bit at a time. The word generator is being used as a 3-bit binary counter that counts from zero (000) to seven (111) at a frequency of 1 kHz. This count is being applied to the multiplexer SELECT input (CBA) and will cause a new bit to be output every 1 millisecond. Therefore, an entire 8-bit input will be transmitted from the multiplexer output in 8 milliseconds. The binary word transmitted will depend on the settings of switches 0–7. The 74151 multiplexer output is being monitored on the logic analyzer screen, with bit 0 being transmitted first and bit 7 transmitted last.

A multiplexer can be wired as a logic function generator to simulate a combinational logic network or a logic gate directly from the truth table without the need for logic simplification. When wired as a logic function generator, the multiplexer SELECT inputs (CBA) are used as the logic network inputs and the multiplexer output (Y) is used as the logic network output. The multiplexer data inputs (D0–D7) are connected high (1) or low (0), as necessary to satisfy the logic network or logic gate truth table. Therefore the logic network or logic gate simulated by the multiplexer can be changed by simply changing the multiplexer data inputs. In order to determine the proper multiplexer data inputs (D0–D7) for simulating a particular logic network or logic gate, a truth table must be plotted. Once the truth table has been established, the multiplexer data inputs (D7–D0) that represent those binary inputs (CBA) that produce a binary one (1) output (Y) on the truth table should be set to binary one (1), and those binary inputs (CBA) that produce a binary zero (0) output (Y) on the truth table should be set to binary zero (0). For example, if binary 000 for inputs CBA produces a binary one (1) for output Y on the truth table, data input D0 on the multiplexer should be high (1).

The circuit in Figure 19-4 is a 74151 multiplexer wired as a 3-input logic gate. The multiplexer SELECT inputs (CBA) represent the logic gate inputs (CBA) and the multiplexer output (Y) represents the logic gate output (Y). The logic gate simulated by the multiplexer depends on the settings of switches 0–7, which control the inputs (D0–D7) on the multiplexer.

Demultiplexing is the opposite process from multiplexing. A demultiplexer accepts digital data from a singular input and connects it to one of many outputs. The demultiplexer input and outputs may be single lines each or multiple lines each. The selection of which output will be connected to the input is determined by the binary code applied to the SELECT inputs. Some demultiplexers have an enable input that may be used to enable or disable it. Similar to a multiplexer, a demultiplexer acts like a digitally controlled multiposition switch.

A decoder can be wired as a demultiplexer if it has an enable input. For this reason, IC chip manufacturers often call these devices decoder/demultiplexers. To use a decoder as a demultiplexer, the decoder inputs are used as the demultiplexer SELECT inputs, one of the decoder enable inputs is used as the demultiplexer input, and the decoder outputs are used as the demultiplexer outputs.

The 74138 decoder/demultiplexer in Figure 19-5 is wired as a demultiplexer. Notice that inputs CBA are being used as the SELECT inputs, the active low enable input G2A is being used as the data input, and the active low outputs Y0–Y7 are being used as the data outputs. Active low enable G2B is connected to ground (0) and active high enable G1 is connected to +5 V (1). When a binary code is applied to the SELECT input (CBA), and a low (0) is applied to input G2A (active low enable), the 74138 is enabled and the code on the SELECT input (CBA) determines which output goes low (0). When a high (1) is applied to input G2A (active low enable), the 74138 is disabled and the outputs all go high (1), including the output selected by the binary code on the SELECT input (CBA). Therefore, the output selected will follow the input on terminal G2A, causing the decoder to act like a demultiplexer.

The circuit in Figure 19-6 is demonstrating a multiplexer/demultiplexer monitoring system. Eight switches are being monitored at a remote location and only four lines are connecting the location of the switches to the location of the monitoring panel. The 74151 multiplexer is monitoring the eight switches and sends a switch position (one or zero) over a single line to the 74138 demultiplexer input (G2A) at the remote monitoring location. The particular switch that is being monitored is determined by the select input (CBA) to the multiplexer and the demultiplexer.

Figure 19-1a Eight-input Multiplexer (EWB Version 4)

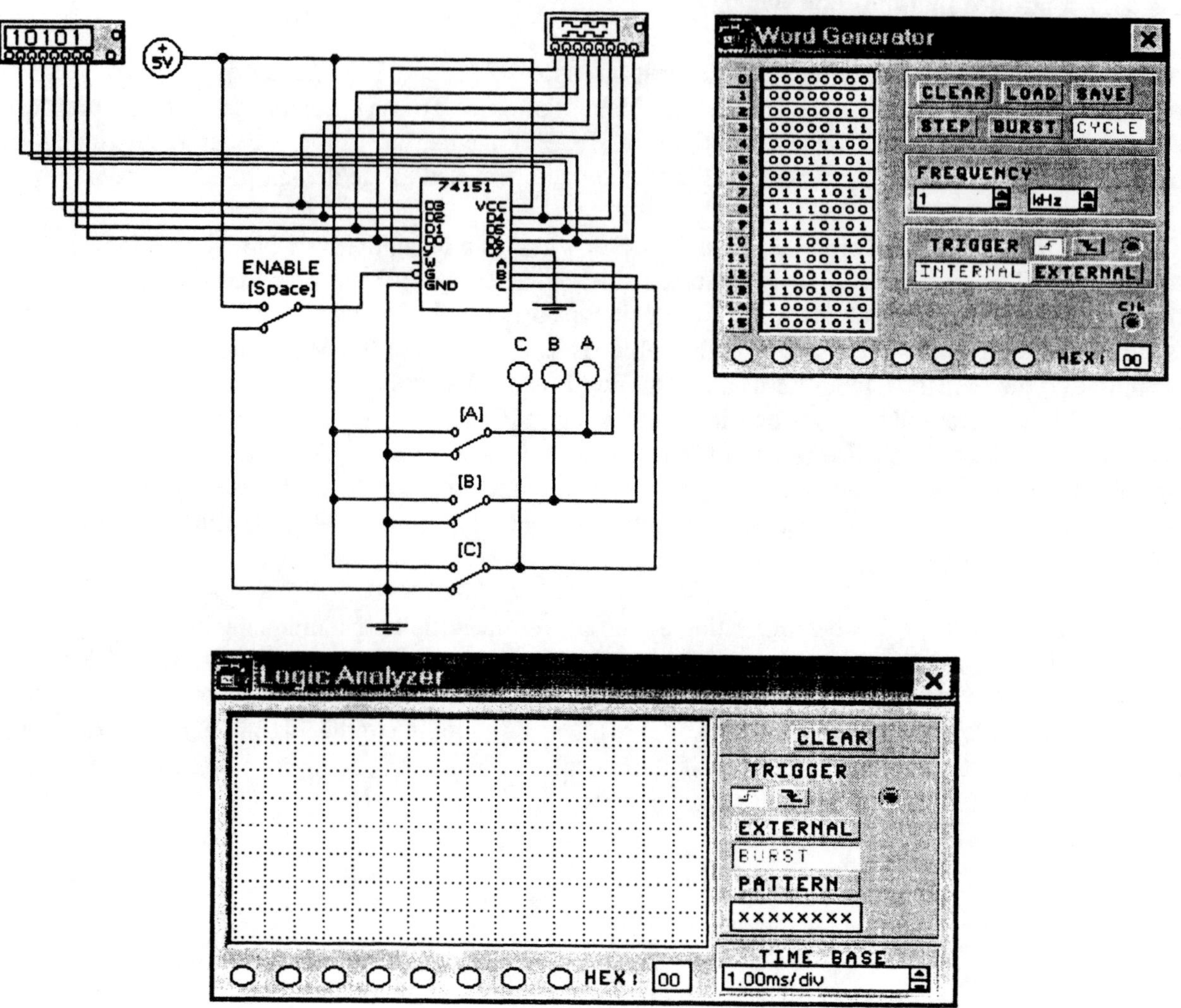

Figure 19-1b Eight-input Multiplexer (EWB Version 5)

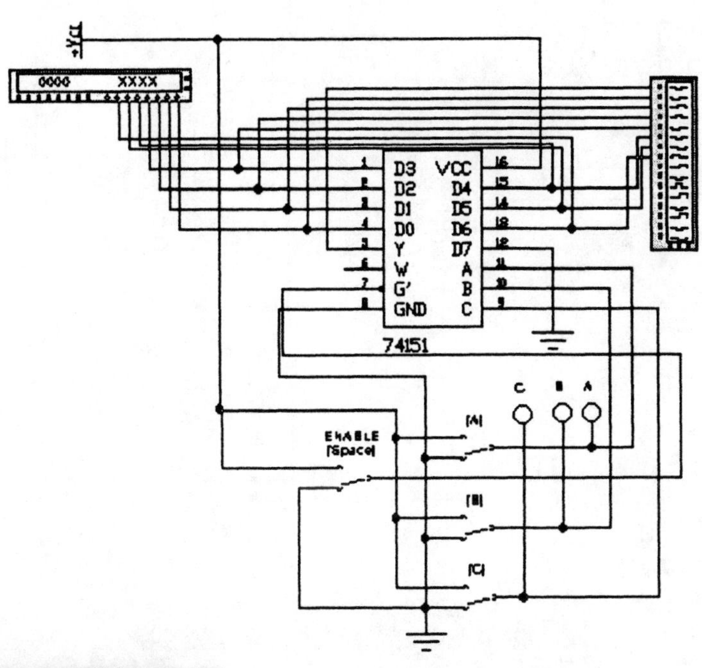

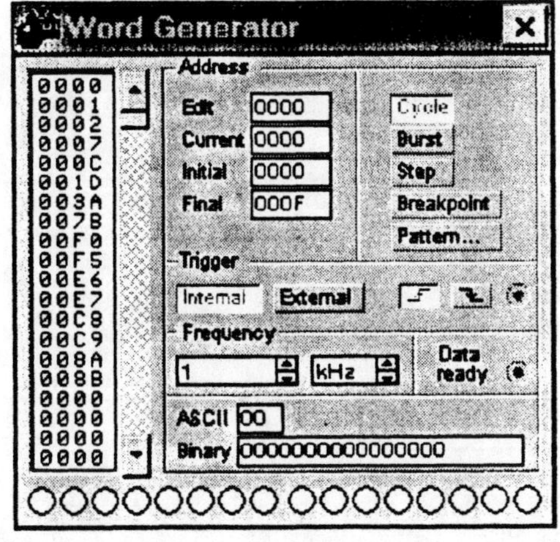

Logic Analyzer Settings
Clocks per division ---------------- 8

Clock Setup dialog box
Clock edge --------------- positive
Clock mode -------------- Internal
Internal clock rate ------ 10 kHz
Clock qualifier ----------- x
Pre-trigger samples --- 100
Post-trigger samples -- 100000
Threshold voltage (V) 2

Trigger Patterns dialog box
A ------------------------------- xxxxxxxxxxxxxxxx
Trigger combinations -- A
Trigger qualifier ---------- x

Figure 19-2a Quad 2-input Multiplexer (EWB Verison 4)

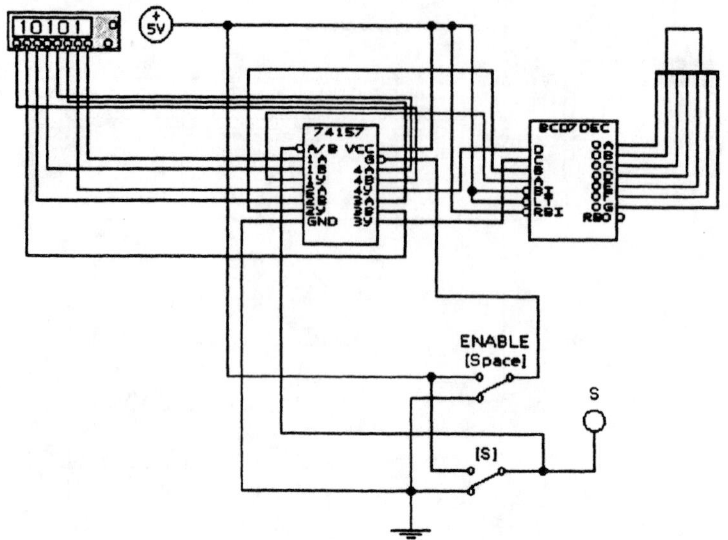

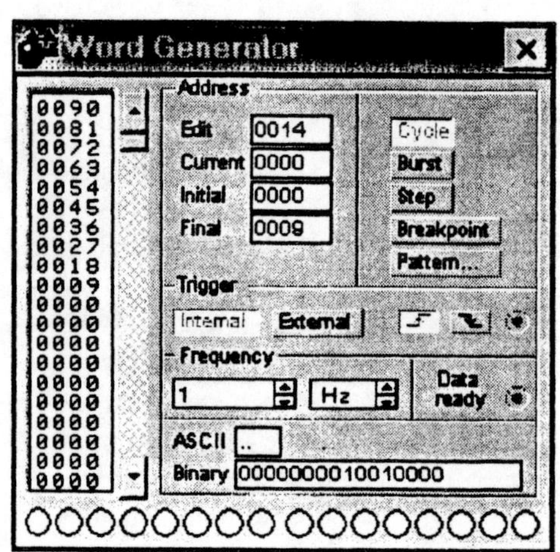

Figure 19-2b Quad 2-input Multiplexer (EWB Version 5)

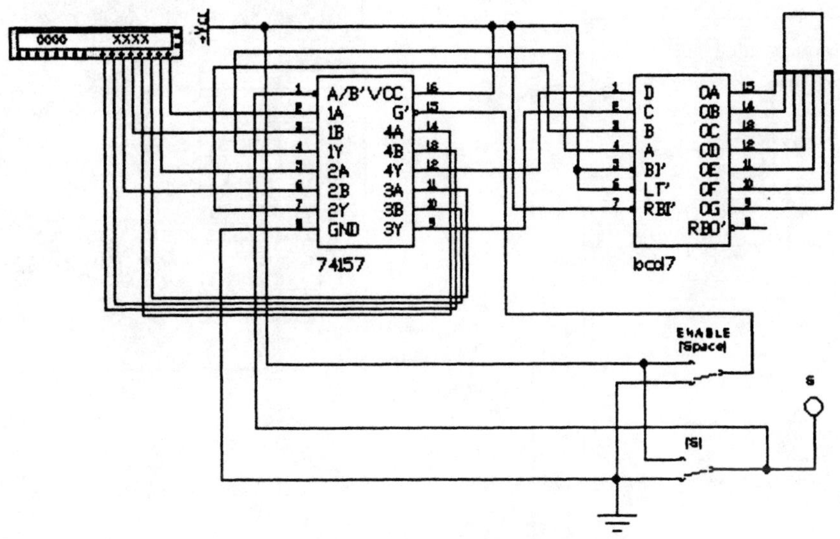

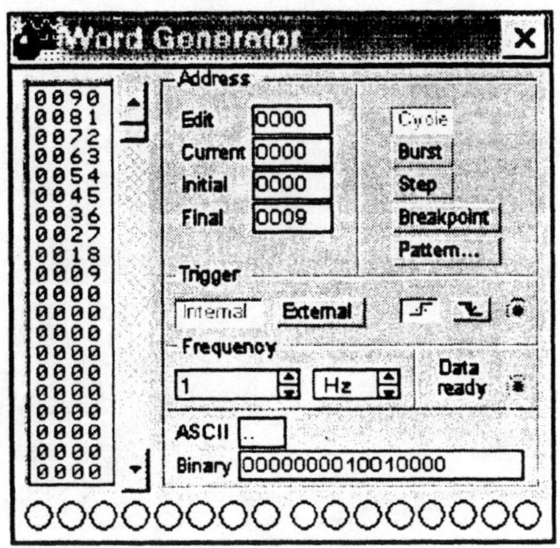

Figure 19-3a Serial Data Transmission (EWB Version 4)

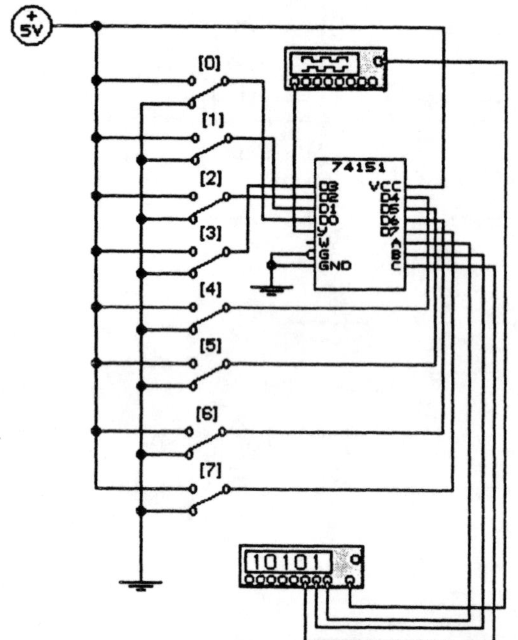

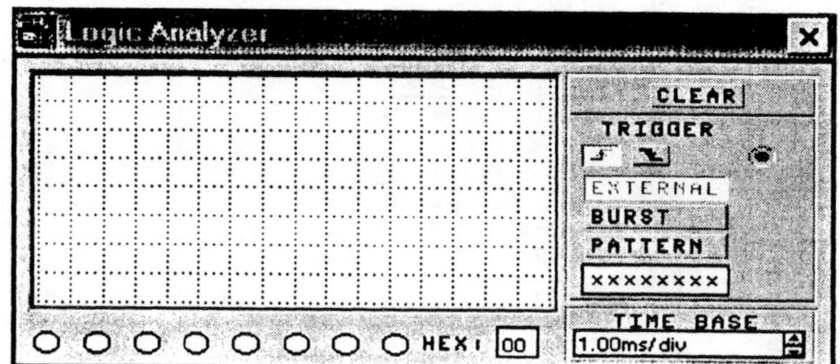

Figure 19-3b Serial Data Transmission (EWB Version 5)

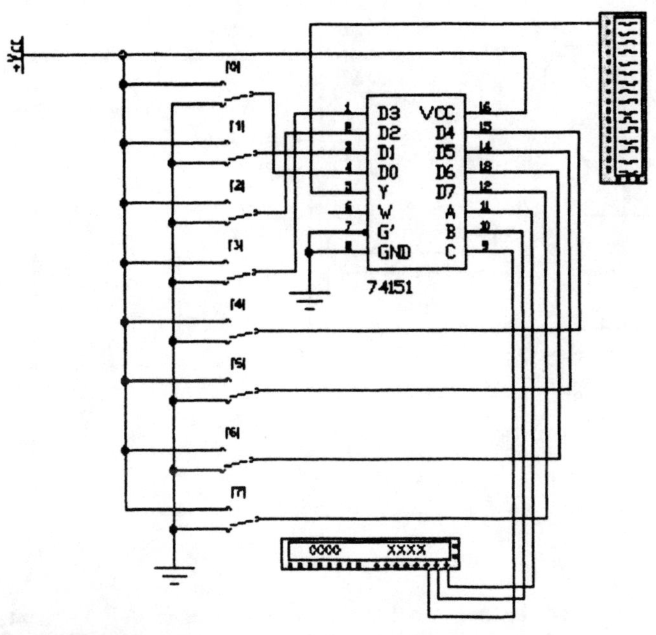

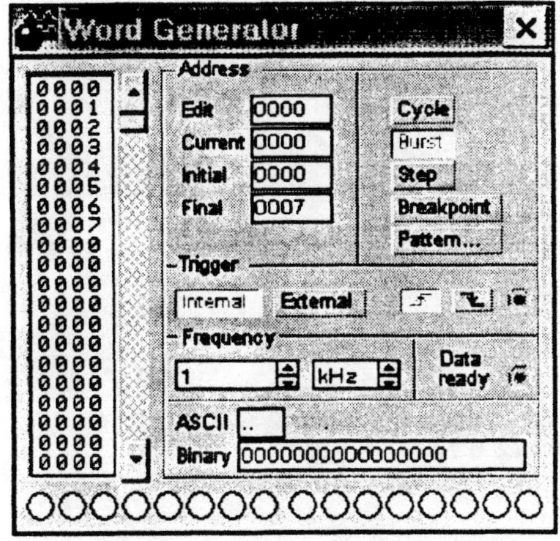

Logic Analyzer Settings
Clocks per division --------------16

Clock Setup dialog box
Clock edge --------------- positive
Clock mode -------------- Internal
Internal clock rate ------ 16 kHz
Clock qualifier ----------- x
Pre-trigger samples --- 100
Post-trigger samples -- 1000
Threshold voltage (V) 2

Trigger Patterns dialog box
A ----------------------------- xxxxxxxxxxxxxxxx
Trigger combinations -- A
Trigger qualifier ---------- x

Figure 19-4 Logic Function Generator

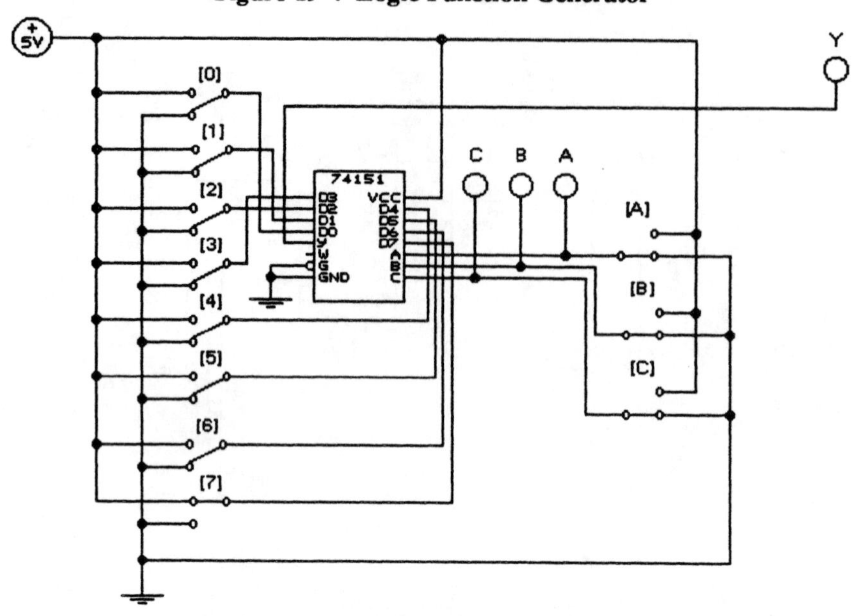

Figure 19-5a 1 Line-to-8 Line Demultiplexer (EWB Version 4)

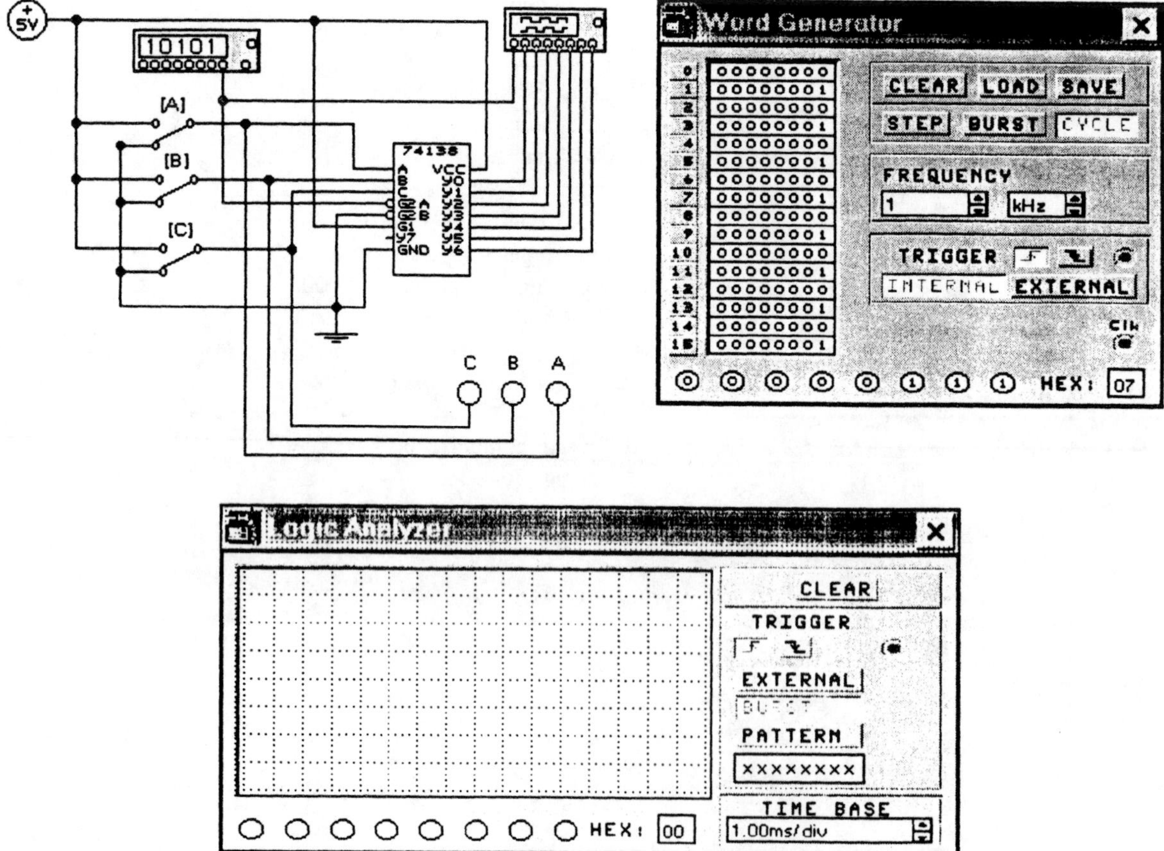

Figure 19-5b 1 Line-to-8 Line Demultiplexer (EWB Version 5)

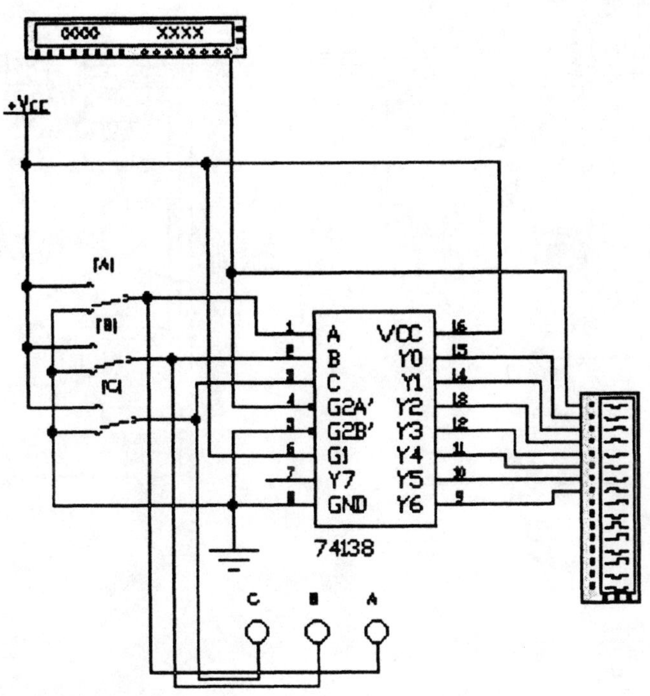

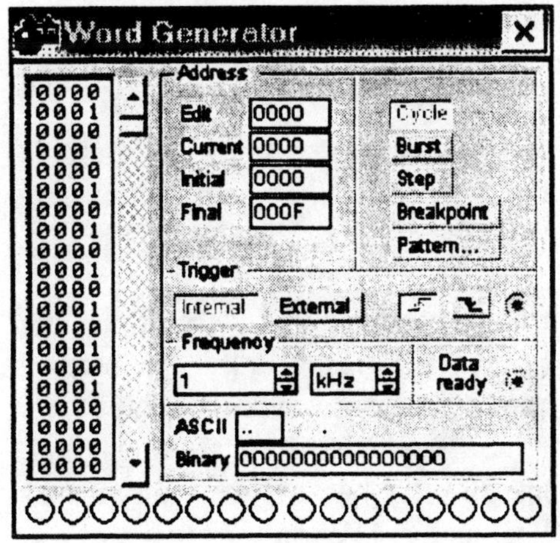

Logic Analyzer Settings
Clocks per division ---------------- 8

Clock Setup dialog box
Clock edge --------------- positive
Clock mode -------------- Internal
Internal clock rate ------ 10 kHz
Clock qualifier ----------- x
Pre-trigger samples --- 100
Post-trigger samples -- 100000
Threshold voltage (V) 2

Trigger Patterns dialog box
A ------------------------------ xxxxxxxxxxxxxxxx
Trigger combinations -- A
Trigger qualifier ---------- x

Figure 19-6 Multiplexer/Demultiplexer Monitoring System

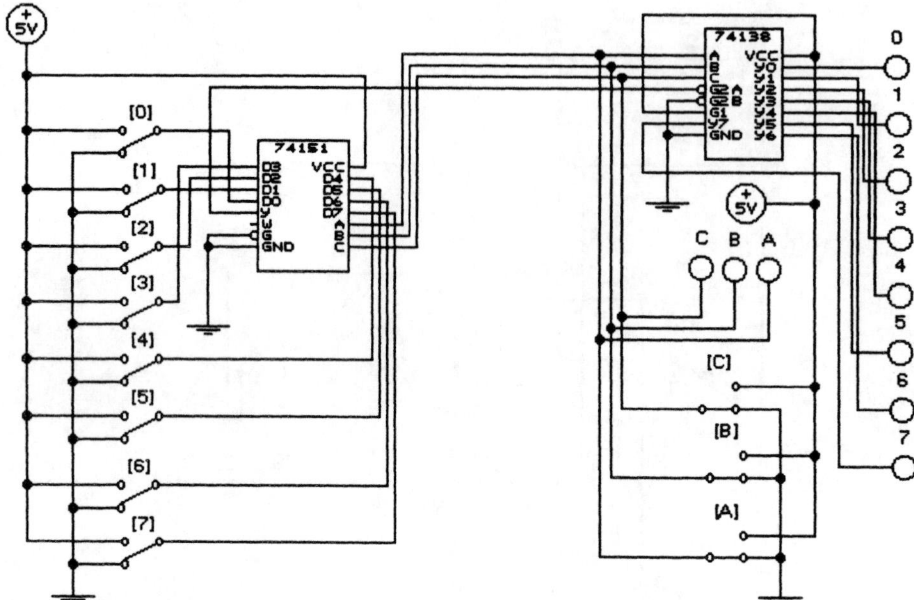

Procedure:

1. Pull down the File menu and open FIG19-1. You are looking at a 74151 8-input multiplexer connected to the word generator, which is supplying a different bit pattern to each multiplexer input (D0–D6). Input D7 is connected to ground (0). The logic analyzer is monitoring the multiplexer output (Y) and inputs D0–D6. The word generator and logic analyzer settings should be as shown in Figure 19-1. Logic switches A, B, and C should be down (0) and the ENABLE switch should be down (G′=0).

2. Click the On-Off switch to run the analysis. Notice that the waveshape at each multiplexer input (D0–D6) is different. The top waveshape (black) is the multiplexer output, the second waveshape from the top (red) is multiplexer input D0, and the bottom waveshape (blue) is multiplexer input D6. Multiplexer input D7 is not being monitored. Notice that the output waveshape (black) is the same as the waveshape for input D0 (red) because the multiplexer SELECT input (CBA) is zero (000). By pressing the A, B, and C keys on the computer keyboard, change the SELECT input (CBA) to values between 000 and 110 and determine which input waveshape (D0–D6) is the same as the output waveshape (black) for each SELECT input. Record your answers in Table 19-1.

NOTE: If you are performing this experiment in a hardwired laboratory, use a 1 kHz pulse generator in place of the word generator. Connect it to each multiplexer input to determine which input the output is following for each SELECT input (CBA). Monitor the multiplexer output with an oscilloscope.

Table 19-1 Eight-input Multiplexer

C	B	A	Input (D0–D6)
0	0	0	
0	0	1	
0	1	0	
0	1	1	
1	0	0	
1	0	1	
1	1	0	

Question: What conclusion can you draw about the relationship between the multiplexer output and the inputs based on the SELECT inputs (CBA)?

3. Set the ENABLE switch to binary one (up) by pressing the space bar on the computer keyboard. This places a binary one (1) on the 74151 multiplexer active low enable terminal ($G'=1$).

Question: What do you notice about the 74151 multiplexer output waveshape (black curve)? Explain your answer.

4. Click the On-Off switch to stop the analysis run. Pull down the File menu and open FIG19-2. You are looking at a 74157 quad 2-input multiplexer wired to a BCD decoder/driver and a 7 segment LED display. The word generator settings should be as shown in Figure 19-2. Select switch S should be down ($S=0$) and the ENABLE switch should be down ($G'=0$). Notice that the low four bits (low nibble) on the word generator are

counting up from zero (0000) to nine (1001) and the high four bits (high nibble) on the word generator are counting down from nine (1001) to zero (0000). The low four bits are connected to the multiplexer A word inputs and the high four bits are connected to the multiplexer B word inputs. The four multiplexer outputs (Y) are connected to the 4-bit input (DCBA) of the BCD-to-7 segment decoder/driver, which will display a decimal number on the 7 segment LED display. Click the On-Off switch to run the analysis.

NOTE: If this part of the experiment is performed in a hardwired laboratory, use a 7448 decoder/driver and a common cathode 7 segment LED display. Follow the IC chip manufacturer's literature for pin numbers and wiring instructions. Use two 4-bit binary counters clocked at a frequency of 1 Hz in place of the word generator.

Questions: Is the 7 segment LED display counting up or down?

Is the 4-bit binary word connected to the multiplexer A input or the 4-bit binary word connected to the B input passing through to the multiplexer output? Explain your answer.

5. Press the S key on the keyboard to set the select input to binary one (1).

Questions: Is the 7 segment LED display counting up or down?

Is the 4-bit binary word connected to the multiplexer A input or the 4-bit binary word connected to the B input passing through to the multiplexer output? Explain your answer.

What conclusion can you draw about the relationship between the multiplexer input words and the output word for each select (S) setting?

6. Press the SPACE bar on the keyboard to set the ENABLE switch to binary one (G′ = 1).

Questions: What happened to the 7 segment LED display?

What conclusion can you draw about the 74157 multiplexer when the ENABLE (G′) input is high (1)?

7. Click the On-Off switch to stop the analysis run. Pull down the File menu and open FIG19-3. You are looking at a 74151 multiplexer wired as a parallel-to-serial converter that transmits a parallel binary word over a single output line one bit at a time. The word generator and logic analyzer settings should be as shown in Figure 19-3. The word generator is being used as a 3-bit binary counter that counts from zero (000) to seven (111) at a frequency of 1 kHz. This count is being applied to the multiplexer SELECT input (CBA) and will cause a new bit to be output every 1 millisecond. Therefore, an entire 8-bit input will be transmitted from the multiplexer output in 8 milliseconds. The binary word transmitted will depend on the settings of switches 0–7. These switches can be changed by pressing the equivalent number key on the computer keyboard. Set the switches (0–7) to any binary word other than zero. Click the On-Off switch to run the analysis. The multiplexer output is being monitored on the logic analyzer screen, with bit 0 being transmitted first and bit 7 transmitted last.

NOTE: If this part of the experiment is being performed in a hardwired laboratory, replace the word generator with a 3-bit binary counter clocked at a frequency of 1 kHz. Replace the logic analyzer with an oscilloscope.

Question: Is the transmitted waveshape on the logic analyzer screen as expected for the switch settings?

8. Click the On-Off switch to stop the analysis. Change the switch settings (0–7). Click the On-Off switch to run the analysis again.

Question: Is the new transmitted waveshape on the logic analyzer screen as expected for the new switch settings?

9. Click the On-Off switch to stop the analysis run. Pull down the File menu and open FIG19-4. You are looking at a 74151 8-input multiplexer wired as a 3-input logic function generator simulating a 3-input logic gate with inputs CBA and output Y. The multiplexer SELECT inputs (CBA) represent the logic gate inputs (CBA) and the multiplexer output (Y) represents the logic gate output (Y). The logic gate simulated by the multiplexer depends on the settings of switches 0–7 (binary inputs to D0–D7 on the multiplexer). Switch 7 should be up (1) and switches 0–6 should be down (0). Click the On-Off switch to run the analysis. By pressing the A, B, and C keys on the keyboard, change logic inputs A, B, and C to the values in Table 19-2 and record the output (Y) for each input combination.

Table 19-2 Truth Table

C	B	A	Y
0	0	0	
0	0	1	
0	1	0	
0	1	1	
1	0	0	
1	0	1	
1	1	0	
1	1	1	

Question: Based on the results in Table 19-2, what 3-input logic gate is being simulated by the multiplexer circuit?

10. Click the On-Off switch to stop the analysis run. Plot the truth table for a 3-input OR gate in Table 19-3.

Table 19-3 Truth Table

C	B	A	Y
0	0	0	
0	0	1	
0	1	0	
0	1	1	
1	0	0	
1	0	1	
1	1	0	
1	1	1	

11. Change switches 0–7 to simulate the OR gate truth table plotted in Table 19-3. Click the On-Off switch to run the analysis. By pressing the A, B, and C keys on the keyboard, test your circuit for all possible input combinations to see if your circuit matches the truth table (Table 19-3).

Question: Did your results indicate that you simulated a 3-input OR gate? If not, what changes are needed?

12. Click the On-Off switch to stop the analysis run. Plot the truth table for a 3-input NAND gate in Table 19-4.

Table 19-4 Truth Table

C	B	A	Y
0	0	0	
0	0	1	
0	1	0	
0	1	1	
1	0	0	
1	0	1	
1	1	0	
1	1	1	

13. Change switches 0–7 to simulate the NAND gate truth table plotted in Table 19-4. Click the On-Off switch to run the analysis. By pressing the A, B, and C keys on the keyboard, test your circuit for all possible input combinations to see if your circuit matches the truth table (Table 19-4).

Question: Did your results indicate that you simulated a 3-input NAND gate? If not, what changes are needed?

14. Click the On-Off switch to stop the analysis run. If you have not completed Experiment 6, complete it now before continuing this experiment.

15. Change switches 0–7 to simulate the logic network truth table plotted for the circuit in Figure 6-4, Experiment 6. Click the On-Off switch to run the analysis. By pressing the A, B, and C keys on the keyboard, test your circuit for all possible input combinations to see if your circuit matches the truth table for the logic circuit in Figure 6-4.

Question: Did your results indicate that you simulated the logic network in Figure 6-4, Experiment 6? If not, what changes are needed?

16. Click the On-Off switch to stop the analysis run. Pull down the File menu and open FIG19-5. You are looking at a 74138 3 line-to-8 line decoder wired as a 1 line-to-8 line demultiplexer. Notice that inputs CBA are being used as the SELECT inputs, the active low enable input G2A is being used as the data input, and the active low outputs Y0–Y7 are being used as the data outputs. The logic analyzer is monitoring outputs Y0–Y6. Output Y7 is not being monitored. Logic switches A, B, and C should be down (0). The word generator and logic analyzer settings should be as shown in Figure 19-5. See the Preparation section for further details.

17. Click the On-Off switch to run the analysis. The top waveshape on the logic analyzer screen is the demultiplexer input (black), the second waveshape from the top (red) is output Y0, and the bottom waveshape (blue) is output Y6. Notice that the output Y0 waveshape (red) is identical to the input waveshape (black) because the SELECT input (CBA) is binary zero (000). By pressing the A, B, and C keys on the keyboard, change the select input (CBA) to values between zero (000) and six (110) and determine which output waveshape is identical to the input waveshape (black). Record your results in Table 19-5.

NOTE: If this part of the experiment is performed in a hardwired laboratory, use a 1 kHz pulse generator in place of the word generator. If a logic analyzer is not available, monitor each demultiplexer output (Y0–Y6) with an oscilloscope to determine which output is active for different SELECT inputs (CBA).

Table 19-5 1 Line-to-8 Line Demultiplexer

C	B	A	Output (Y0–Y6)
0	0	0	
0	0	1	
0	1	0	
0	1	1	
1	0	0	
1	0	1	
1	1	0	

Question: What conclusion can you draw about the relationship between the demultiplexer outputs and the input, based on the SELECT inputs (CBA)?

18. Click the On-Off switch to stop the analysis run. Pull down the File menu and open FIG19-6. You are looking at a multiplexer/demultiplexer monitoring system that will demonstrate an application for a multiplexer and a demultiplexer. Eight switches are being monitored at a remote location without running eight separate lines between the switches and the remote monitoring panel. For a more detailed discussion of the system, see the Preparation section.

19. Click the On-Off switch run the analysis. Press the number keys on the computer keyboard to switch the numbered switches on and off (up and down) to determine which switch is being monitored by the logic probe lights.

Question: Which switch is being monitored by the logic probe lights? Explain why that particular switch is being monitored.

20. Change the SELECT input (CBA) to another binary code by pressing the A, B, and C keys on the keyboard. Press the number keys again to switch the numbered switches on and off (up and down) to determine which switch is being monitored by the logic probe lights.

Question: Is the correct switch being monitored by the logic probe lights?

Name ______________________

Date ______________________

Troubleshooting MSI Logic Circuits

Objectives:

1. Determine the defective component in an encoder/decoder decimal display system.
2. Determine the defective component in a multiplexer/decoder decimal display system.
3. Determine the defective component in a multiplexer/demultiplexer security monitoring system.

Materials:

This experiment can only be performed on Electronics Workbench using the circuits disk provided with this manual.

Preparation:

In order to perform this experiment effectively, you must first complete Experiments 18 and 19. Use the theory learned in those experiments to find the defective components in the logic circuits in this experiment.

Before attempting to find the defective component, test the circuit for all possible input conditions to determine which input conditions are producing the incorrect outputs. This should give you some direction in determining where to begin testing. While the input is producing an incorrect output, use the logic probe light to measure logic levels at various test points to help determine the defective component.

Procedure:

1. Pull down the File menu and open FIG20-1. Click the On-Off switch to run the analysis. This circuit should display the decimal number of the highest numbered switch that is down (0) on the LED display. Using the logic probe light to measure the logic levels at various test points in the circuit, determine which component is defective. To switch the numbered switches, press the key on the computer keyboard that matches the number label on the switch.

   Defective component ______________

2. Pull down the File menu and open FIG20-2. Click the On-Off switch to run the analysis. This circuit should display the decimal number of the highest numbered switch that is down (0) on the LED display. Using the logic probe light to measure the logic levels at various test points in the circuit, determine which component is defective. To switch the numbered switches, press the key on the computer keyboard that matches the number label on the switch.

   Defective component ________________

3. Pull down the File menu and open FIG20-3. Click the On-Off switch to run the analysis. This circuit should display the decimal number of the highest numbered switch that is down (0) on the LED display. Using the logic probe light to measure the logic levels at various test points in the circuit, determine which component is defective. To switch the numbered switches, press the key on the computer keyboard that matches the number label on the switch.

   Defective component ________________

4. Pull down the File menu and open FIG20-4. Click the On-Off switch to run the analysis. This circuit should display a decimal number counting up (A word) when S=0 or counting down (B word) when S=1. Using the logic probe light to measure the logic levels at various test points in the circuit, determine which component is defective.

   Defective component ________________

5. Pull down the File menu and open FIG20-5. Click the On-Off switch to run the analysis. This circuit should display a decimal number counting up (A word) when S=0 or counting down (B word) when S=1. Using the logic probe light to measure the logic levels at various test points in the circuit, determine which component is defective.

   Defective component ________________

6. Pull down the File menu and open FIG20-6. Click the On-Off switch to run the analysis. The code on the SELECT input (CBA) should determine which switch is being monitored. The remaining logic lights should be "on" (1) continuously. To switch the switches, press the key on the computer keyboard that matches the number or letter label on the switch. Using the logic probe light to measure the logic levels at various test points in the circuit, determine which component is defective.

   Defective component ________________

7. Pull down the File menu and open FIG20-7. Click the On-Off switch to run the analysis. The code on the SELECT input (CBA) should determine which switch is being monitored. The remaining logic lights should be "on" (1) continuously. To switch the switches, press the key on the computer keyboard that matches the number or letter label on the switch. Using the

logic probe light to measure the logic levels at various test points in the circuit, determine which component is defective.

Defective component ______________

8. Pull down the File menu and open FIG20-8. Click the On-Off switch to run the analysis. The code on the SELECT input (CBA) should determine which switch is being monitored. The remaining logic lights should be "on" (1) continuously. To switch the switches, press the key on the computer keyboard that matches the number or letter label on the switch. Using the logic probe light to measure the logic levels at various test points in the circuit, determine which component is defective.

Defective component ______________

Part

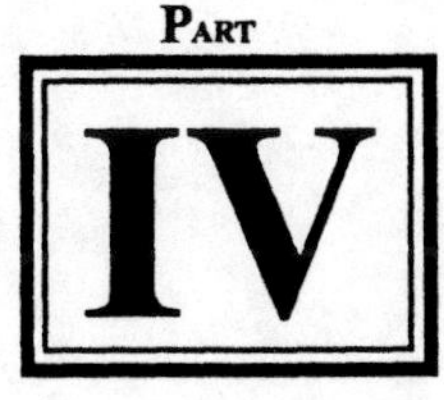

Sequential Logic Circuits

The logic circuits studied thus far have been combinational logic circuits, which have output responses that follow changes in the input levels with minimum time delay. Prior input conditions have no effect on the present output levels because combinational logic circuits do not have memory.

The experiments in Part IV involve the study of sequential logic circuits. Sequential logic circuits have present outputs that are affected by prior input levels because they have memory. The basic building block of sequential logic circuits is the flip-flop or latching circuit. In the first two experiments, you will study how latches and flip-flops store binary data. Next, you will study how monostable and astable multivibrators (pulse oscillators) generate pulses. In the remaining experiments, you will study registers and counters and learn how they are wired using basic flip-flops. In the final experiment you will troubleshoot some sequential logic circuits involving latches, flip-flops, registers, and counters.

The circuits for the experiments in Part IV can be found on the enclosed disk in the PART4 subdirectory.

Name ______________________

Date ______________________

S-R and D Latches

Objectives:

1. Investigate the operation of a NOR gate S-R latch.
2. Investigate the operation of a NAND gate S-R latch.
3. Investigate the operation of a D latch.

Materials:

One 5 V dc power supply
Two logic switches
Four logic probe lights
Two two-input NOR gates (1-7402 IC)
Four two-input NAND gates (1-7400 IC)
One INVERTER (1-7404 IC)
One D latch (1-7475 IC)
Two square wave generators
One logic analyzer or dual-trace oscilloscope

Preparation:

A latch is a bistable storage device with an output that can be latched into one of two states. This is caused by a feedback arrangement in which the outputs of two logic gates are fed back to the opposite gate inputs.

The most basic binary data storage device is the set-reset (S-R) latch. The S-R latch has two inputs, set (S) and reset (R). The output (Q) will latch high (1) when the set (S) input is active and will latch low (0) when the reset (R) input is active.

A D latch has only one input. The D latch also has an enable (EN) terminal, which is used to control whether the latch accepts or ignores the input (D). When the latch is enabled, the output (Q) will follow the input (D). When the latch is disabled, the output (Q) will "latch" (store) the last input level (D) before it was disabled. Because the output (Q) follows the input (D) when the latch is enabled, D latches are often referred to as "transparent" latches.

The logic circuit in Figure 21-1 is a set-reset (S-R) latch wired with two NOR gates and is called a NOR gate S-R latch. When the set (S) input is high (1) and the reset (R) input is low (0), the output (Q) will become latched in the high state (Q=1). When both the set (S) and reset (R) inputs are low (0), the output (Q) stays latched at its previous state. When the reset (R) input is high (1) and the set (S) input is low (0), the output (Q) will become latched in the low state (Q=0). The Q' output will always be the inverse of the Q output, except when both the set (S) and reset (R) inputs are high (1), which is a "not allowed" input condition for a NOR gate latch.

The logic circuit in Figure 21-2 is a set-reset (S-R) latch wired with two NAND gates and is called a NAND gate S-R latch. When the set (S') input is low (0) and the reset (R') input is high (1), the output (Q) will become latched in the high state (Q=1). When both the set (S') and reset (R') inputs are high (1), the output (Q) stays latched at its previous state. When the reset (R') input is low (0) and the set (S') input is high (1), the output (Q) will become latched in the low state (Q=0). Therefore, the set (S') and reset (R') inputs of a NAND latch are active low inputs. The Q' output will always be the inverse of the Q output, except when both the set (S') and reset (R') inputs are low (0), which is a "not allowed" input condition for a NAND gate latch.

The logic circuit in Figure 21-3 is a D latch wired with four 2-input NAND gates and an INVERTER. When the enable input (EN) is low (0), latch inputs S' and R' are both high (1). This causes the S-R latch to stay "latched" on the previous input and not respond to any D inputs. When the enable input (EN) is high (1), latch inputs S' and R' depend on whether input D is high (1) or low (0). If input D is high (1), latch input S' is low (0) and input R' is high (1), causing the latch to set and the output to go high (Q=1). If input D is low (0), latch input S' is high (1) and input R' is low (0), causing the latch to reset and the output to go low (Q=0). Therefore, the latch output (Q) will follow the input (D) when the enable (EN) terminal is high (1) and the latch output (Q) will stay "latched" when the enable (EN) terminal is low (0).

The 7475 in Figure 21-4 has four D latches that are similar to the D latch in Figure 21-3. The enable terminals for latches 1 and 2 are connected together (1C,2C) and the enable terminals for latches 3 and 4 are connected together (3C,4C). In this experiment, you will only use latch 1. When the enable terminal (1C,2C) is high (1), the output (Q) will follow the input (D) and the latch is "transparent." When the enable terminal (1C,2C) is low (0), the output (Q) is "latched."

The circuit in Figure 21-5 will monitor the timing of the 7475 D latch. The word generator will apply a pulse waveshape to the D input and a different pulse waveshape to the enable input (EN). The logic analyzer will monitor the D input, the enable input (EN), and the output (Q). When the enable input (EN) is high (1), the D latch is "transparent" and the output (Q) will follow the input (D). When the enable input (EN) is low (0), the D latch output is "latched" on the value of input D when the enable went low (0).

Figure 21-1 NOR Gate S-R Latch

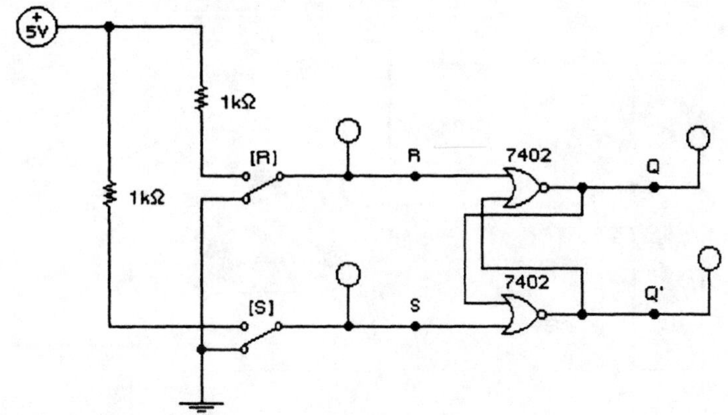

Figure 21-2 NAND Gate S-R Latch

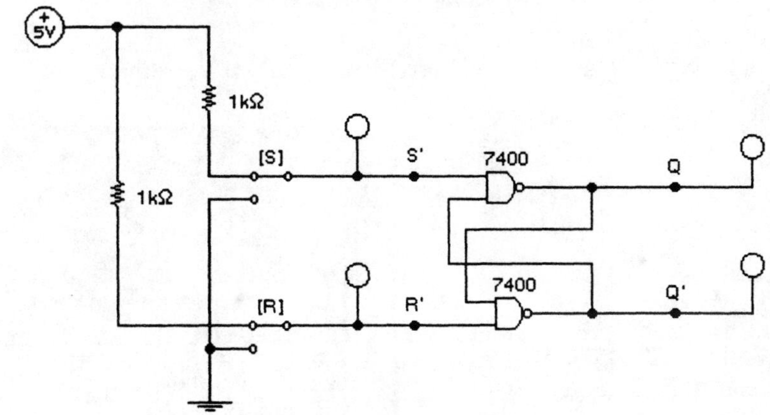

Figure 21-3 NAND Gate D Latch

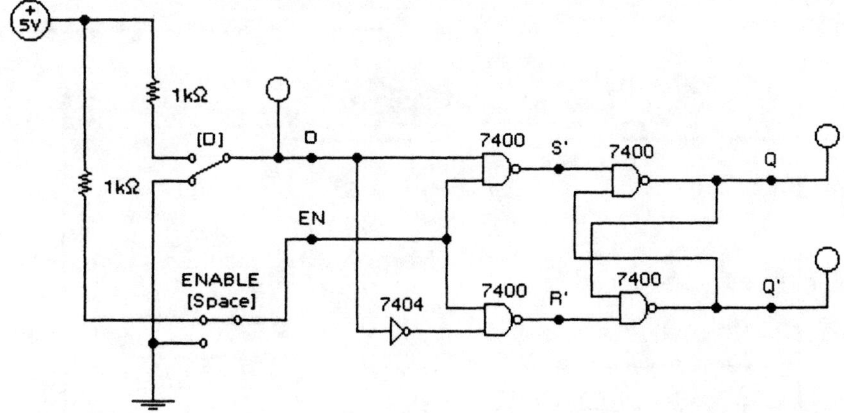

Figure 21-4 7475 D Latch

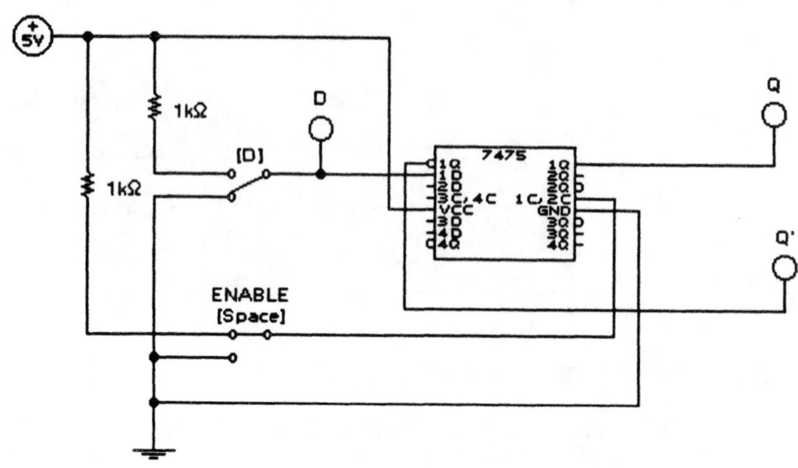

Figure 21-5a 7475 D Latch Timing (EWB Version 4)

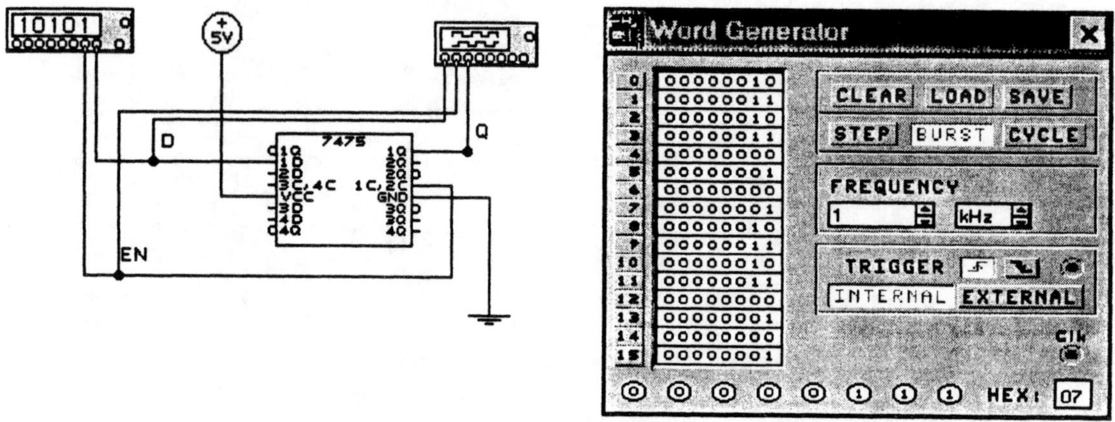

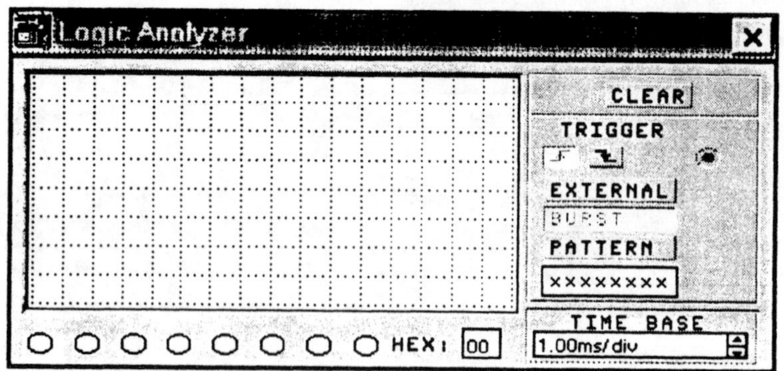

Figure 21-5b 7475 D Latch Timing (EWB Version 5)

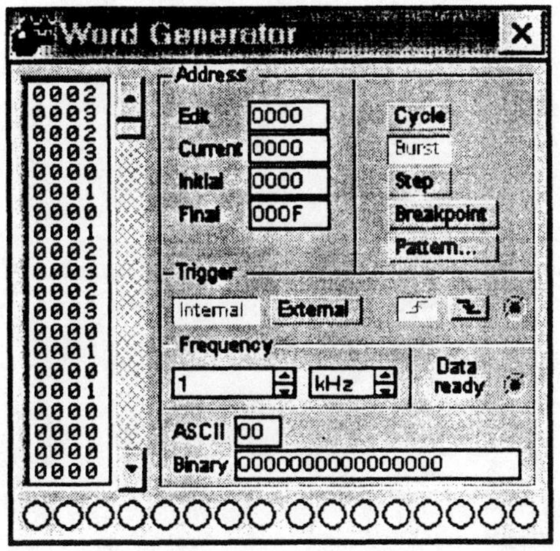

Logic Analyzer Settings
Clocks per division ---------------- 8

Clock Setup dialog box
Clock edge --------------- positive
Clock mode -------------- Internal
Internal clock rate ------ 10 kHz
Clock qualifier ----------- x
Pre-trigger samples --- 100
Post-trigger samples -- 1000
Threshold voltage (V) 2

Trigger Patterns dialog box
A ------------------------------ xxxxxxxxxxxxxxxx
Trigger combinations -- A
Trigger qualifier ---------- x

Procedure:

1. Pull down the File menu and open FIG21-1. You are looking at an S-R latch wired with two 2-input NOR gates. Both the S and the R switches should be down (0). Click the On-Off switch to run the analysis. Press the S key on the computer keyboard to bring the set (S) input to binary one (1).

NOTE: When this experiment is performed in a hardwired laboratory, the output (Q) will set (1) or reset (0) when power is turned on. This does not happen in the computer simulation because there is no power-up transient to trigger the latch.

Question: What are outputs Q and Q′ when S=1 and R=0?

2. Press the S key again to bring the set (S) input to binary zero (0).

Question: What happened to outputs Q and Q' when S=0 and R=0?

3. Press the R key to bring the reset (R) input to binary one (1).

Question: What are outputs Q and Q' when S=0 and R=1?

4. Press the R key again to bring the reset (R) input to binary zero (0).

Question: What happened to outputs Q and Q' when S=0 and R=0?

5. Press the S key and the R key to bring both the set (S) and reset (R) inputs to binary one (1).

Question: What happened to outputs Q and Q' when S=1 and R=1? Explain.

6. Click the On-Off switch to stop the analysis run. Pull down the File menu and open FIG21-2. You are looking at an S-R latch wired with two 2-input NAND gates. Both the S and the R switches should be up (1). Click the On-Off switch to run the analysis. Press the S key on the computer keyboard to bring the set (S') input to binary zero (0).

NOTE: When this experiment is performed in a hardwired laboratory, the output (Q) will set (1) or reset (0) when power is turned on. This does not happen in the computer simulation because there is no power-up transient to trigger the latch.

Question: What are outputs Q and Q' when S'=0 and R'=1?

7. Press the S key again to bring the set (S') input to binary one (1).

Question: What happened to outputs Q and Q' when S'=1 and R'=1?

8. Press the R key to bring the reset (R') input to binary zero (0).

Question: What are outputs Q and Q' when S'=1 and R'=0?

9. Press the R key again to bring the reset (R') input to binary one (1).

Question: What happened to outputs Q and Q' when S'=1 and R'=1?

10. Press the S key and the R key to bring both the set (S') and reset (R') inputs to binary zero (0).

Question: What happened to outputs Q and Q' when S'=0 and R'=0? Explain.

11. Click the On-Off switch to stop the analysis run. Pull down the File menu and open FIG21-3. You are looking at a D latch wired with four 2-input NAND gates and an INVERTER. See the Preparation section for a detailed discussion of this logic network. Switch D should be down (0) and the ENABLE switch should be up (1). Click the On-Off switch to run the analysis.

Question: What are outputs Q and Q' with D=0?

12. Press the D key to set the D input to binary one (1).

Question: What happened to outputs Q and Q' with D=1?

13. Press the D key again to clear the D input to binary zero (0).

Question: What happened to output Q? Explain.

14. Press the space bar to bring the enable (EN) terminal to binary zero (0). Press the D key to set the D input to binary one (1).

Question: What happened to output Q? Explain.

15. Click the On-Off switch to stop the analysis run. Pull down the File menu and open FIG21-4. You will test a 7475 D latch and compare the results with the results for the NAND gate D

latch in Figure 21-3. Switch D should be down (0) and the ENABLE switch should be up (1). Click the On-Off switch to run the analysis.

Questions: What is output Q with D=0?

How do your results compare with the results for the NAND gate D latch in Figure 21-3?

16. Press the D key to set the D input to binary one (1).

Questions: What is output Q with D=1?

How do your results compare with the results for the NAND gate D latch in Figure 21-3?

17. Press the space bar to bring the enable terminal (1C,2C) to binary zero (0). Press the D key to clear the D input to binary zero (0).

Questions: What happened to output Q? Explain.

How do your results compare with the results for the NAND gate D latch in Figure 21-3?

18. Click the On-Off switch to stop the analysis run. Pull down the File menu and open FIG21-5. You are looking at a circuit for measuring the timing of the 7475 D latch. The word generator and logic analyzer settings should be as shown in Figure 21-5. Click the On-Off switch to run the analysis. The word generator is applying a pulse waveshape to the D input and a different pulse waveshape to the enable input (EN) of the D latch. The logic analyzer is monitoring input D (red), enable input EN (green), and output Q (blue).

NOTE: If this experiment is performed in a hardwired laboratory, use two pulse generators in place of the word generator and a dual-trace oscilloscope in place of the logic analyzer if a logic analyzer is not available.

Questions: What do you notice about output Q (blue) when enable EN (green) is high (1)?

What do you notice about output Q (blue) when enable EN (green) is low(0)?

Name ______________________
Date ______________________

Edge-Triggered Flip-Flops

Objectives:

1. Investigate the operation of an edge-triggered S-R flip-flop.
2. Investigate the operation of an edge-triggered D flip-flop.
3. Investigate the operation of an edge-triggered J-K flip-flop.

Materials:

One 5 V dc power supply
Five logic switches
Four logic probe lights
Two two-input NOR gates (1-7402 IC)
Four two-input AND gates (1-7408 IC)
Three INVERTERS (1-7404 IC)
One positive-edge-triggered D flip-flop (1-7474 IC)
Two three-input NAND gates (1-7410 IC)
One negative-edge-triggered J-K flip-flop (1-74112 IC)
Three square wave generators
One logic analyzer or dual-trace oscilloscope

Preparation:

Logic circuits can be either asynchronous or synchronous. In an asynchronous logic circuit, the output changes state any time an input changes state. In a synchronous logic circuit, the exact time at which the output can change state is determined by a series of rectangular pulses called clock pulses. A flip-flop is a synchronous bistable device. An edge-triggered flip-flop output changes state only on the positive edge (rising edge) or only on the negative edge (falling edge) of the clock pulse, depending on whether it is a positive-edge-triggered or a negative-edge-triggered flip-flop.

The logic circuit in Figure 22-1 is an S-R latch wired as a positive-edge-triggered S-R flip-flop. The INVERTER (IC1) and the AND gate (IC2A) form an edge detector circuit that provides a short duration pulse at the edge detector output (EN) on the positive edge of the clock pulse at the edge detector input (CLK). The circuit is able to produce the short duration pulse at the output (EN) because there is a short (few nanoseconds) time delay between the AND gate input from the

INVERTER and the AND gate input from the clock (CLK). Therefore, when the clock pulse rises to binary one (1), the output of the INVERTER is still one (1) for a few nanoseconds while the other AND gate input goes to one (1). This produces the short duration one (1) output at terminal EN. This will open the pulse steering circuit (IC2B and IC2C) for a short duration (few nanoseconds), allowing the S-R latch to set (Q=1), clear (Q=0), or stay the same based on the inputs at S and R during the clock transition period. When inputs S=1 and R=0, the flip-flop output will set (Q=1) on the positive edge of the clock pulse. When inputs S=0 and R=1, the flip-flop output will clear (Q=0) on the positive edge of the clock pulse. When inputs S=0 and R=0, the flip-flop output will not change state on the positive edge of the clock pulse. Inputs S=1 and R=1 are "not allowed" input conditions for an S-R flip-flop. The edge detector circuit can be changed to produce a short duration pulse on the negative edge of the clock pulse by placing an INVERTER at the clock input (CLK). This would produce a negative-edge-triggered flip-flop.

The logic circuit in Figure 22-2 is an S-R latch wired as a positive-edge-triggered D flip-flop. It is identical to the positive-edge-triggered S-R flip-flop in Figure 22-1, except an INVERTER (IC1B) is connected between the S and R inputs of the S-R flip-flop to make the D flip-flop. If D=1, the flip-flop output will set (Q=1) when the clock input (CLK) receives a positive clock edge. If D=0, the flip-flop output will clear (Q=0) when the clock input (CLK) receives a positive clock edge. The positive-edge-triggered D flip-flop will not respond to changes in the D input until the clock input (CLK) receives a positive clock edge. A negative-edge-triggered D flip-flop operates the same as a positive-edge-triggered D flip-flop except it responds to a negative clock edge instead of a positive clock edge. The D flip-flop is often used to store a single data bit.

The circuit in Figure 22-3 will display the timing of the positive-edge-triggered D flip-flop. The word generator is applying a series of clock pulses to the clock input (CLK) and a pulse waveshape to the D input of the flip-flop. The logic analyzer is monitoring the CLK input, the output of the edge detector circuit (EN), the D input, the output (Q), and the inverted output (Q'). The output of the flip-flop circuit will set (Q=1) on the positive edge of the clock pulse when D=1, and clear (Q=0) on the positive edge of the clock pulse when D=0. The flip-flop inverted output (Q') will be the inverse of the output (Q).

The 7474 in Figure 22-4 has two positive-edge-triggered D flip-flops with active low asynchronous preset (PRE) and clear (CLR) inputs. You will use only one of the D flip-flops in this experiment. The asynchronous inputs override the CLK (synchronous) input and will set (Q=1) or clear (Q=0) the flip-flop output whenever a binary zero (0) is applied to the PRE or CLR active low terminals. These asynchronous inputs are not edge-triggered and respond to dc levels. The S and C switches control the voltage levels (high or low) applied to the asynchronous PRE and CLR inputs. Setting both asynchronous PRE and CLR terminals low (0) at the same time is a "not allowed" input condition.

The circuit in Figure 22-5 will display the timing of the 7474 positive-edge-triggered D flip-flop. The word generator is applying a series of clock pulses to the clock input (CLK) and a pulse waveshape to the D input of the flip-flop. The logic analyzer is monitoring the CLK input, the D input, the output (Q), and the inverted output (Q'). The output of the flip-flop circuit will set (Q=1) on the positive edge of the clock pulse when D=1, and clear (Q=0) on the positive edge of the clock pulse when D=0. The flip-flop inverted output (Q') will be the inverse of the output (Q).

The logic circuit in Figure 22-6 is an S-R latch wired as a positive-edge-triggered J-K flip-flop. The J-K flip-flop is similar to the S-R flip-flop, except the J input is the set input and the K input is the reset (clear) input. When inputs J=1 and K=0, the flip-flop output will set (Q=1) on the clock edge. When inputs J=0 and K=1, the flip-flop output will clear (Q=0) on the clock edge. When inputs J=0 and K=0, the flip-flop output (Q) will not change state on the clock edge. The most important feature of the J-K flip-flop is an additional state with inputs J=1 and K=1, which is "not allowed" in the S-R flip-flop. When inputs J=1 and K=1, the J-K flip-flop output will toggle between set (Q=1) and clear (Q=0) on each clock edge. For a detailed discussion of how the J-K flip-flop toggles, see your digital electronics textbook. The J-K flip-flop is the most versatile and widely used type of flip-flop.

The 74112 in Figure 22-7 has two negative-edge-triggered J-K flip-flops with active low asynchronous preset (PRE) and clear (CLR) inputs. You will use only one of the J-K flip-flops in this experiment. The asynchronous inputs on the 74112 operate the same as the asynchronous inputs on the 7474 D flip-flop. (See the discussion of the 7474 D flip-flop). The S and C switches control the voltage levels (high or low) applied to the asynchronous PRE and CLR inputs.

The circuit in Figure 22-8 will display the timing of the 74112 negative-edge-triggered J-K flip-flop. The word generator is applying a series of clock pulses to the clock input (CLK), a pulse waveshape to the J input, and a different pulse waveshape to the K input of the flip-flop. The logic analyzer is monitoring the CLK input, the J input, the K input, the output (Q), and the inverted output (Q'). When J=1 and K=0, the output of the flip-flop will set (Q=1) on a negative edge of the clock pulse. When J=0 and K=1, the output will clear (Q=0) on a negative edge of the clock pulse. When J=1 and K=1, the output will toggle on a negative edge of the clock pulse. When J=0 and K=0, the output will not change on a negative edge of the clock pulse. The flip-flop inverted output (Q') will be the inverse of the output (Q).

Figure 22-1 Edge-Triggered S-R Flip-Flop

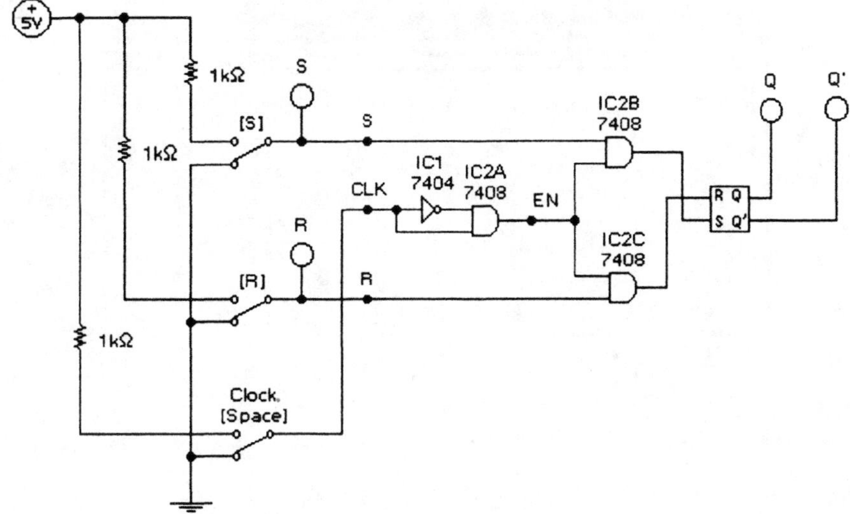

Figure 22-2 Edge-Triggered D Flip-Flop

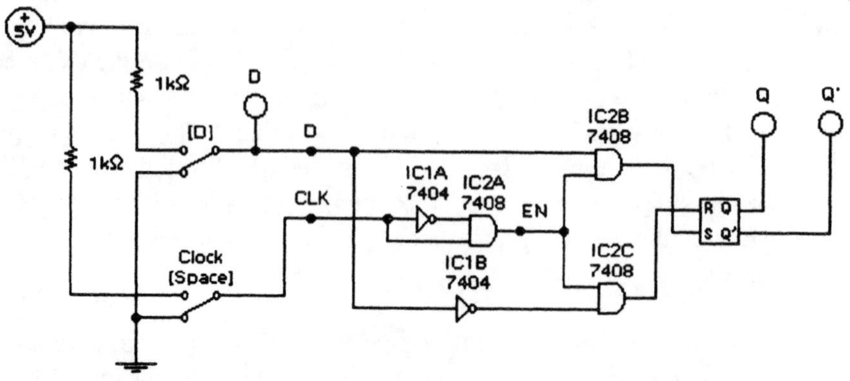

Figure 22-3a Edge-Triggered D Flip-Flop Timing (EWB Version 4)

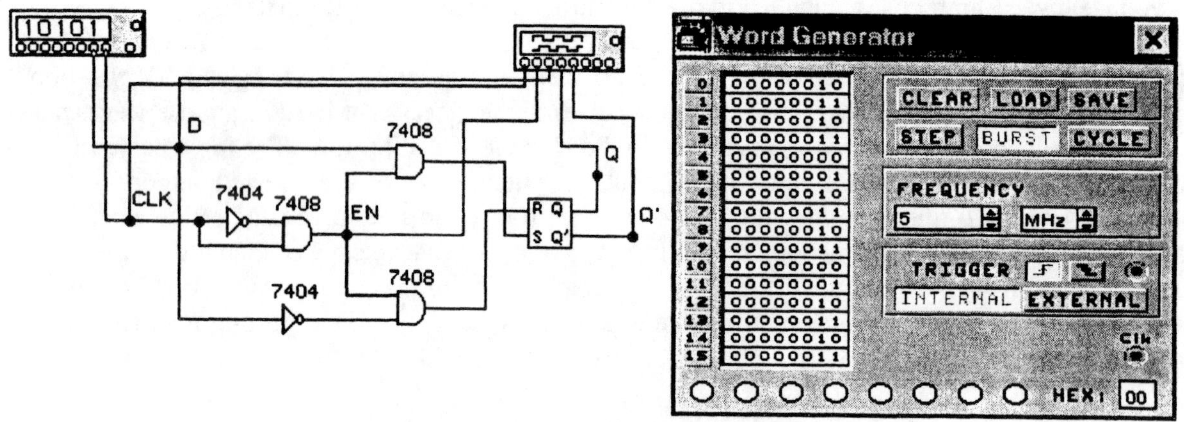

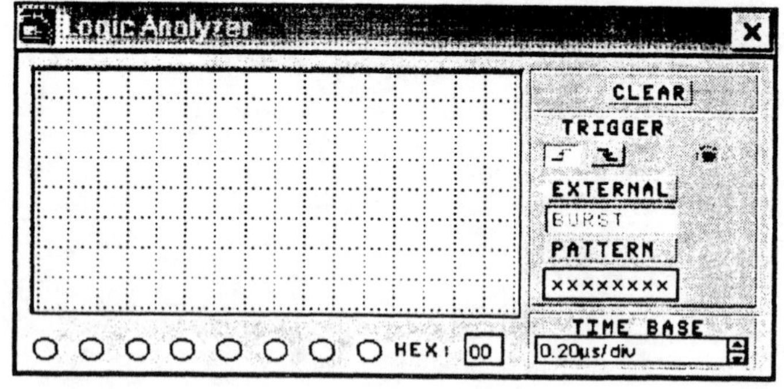

Figure 22-3b Edge-Triggered D Flip-Flop Timing (EWB Version 5)

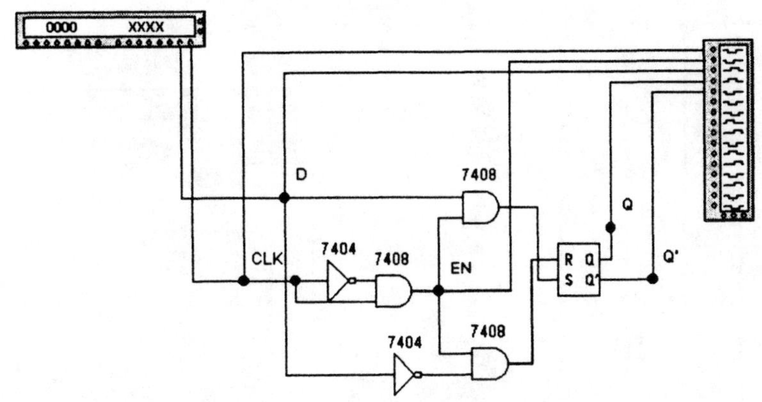

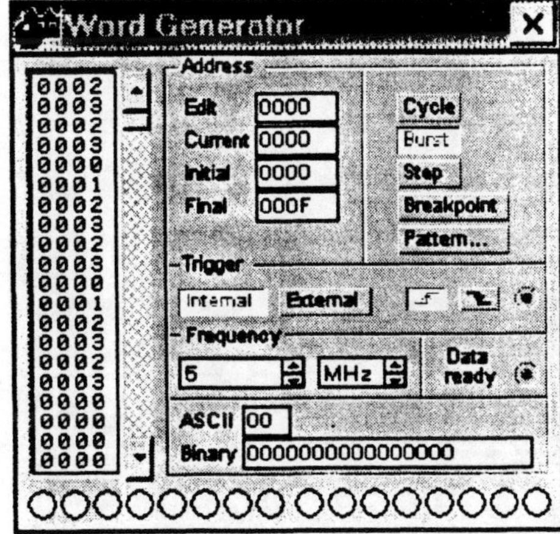

Logic Analyzer Settings
Clocks per division ---------------- 8

Clock Setup dialog box
Clock edge --------------- positive
Clock mode -------------- Internal
Internal clock rate ------ 50 MHz
Clock qualifier ----------- x
Pre-trigger samples --- 0
Post-trigger samples -- 1000
Threshold voltage (V) 2

Trigger Patterns dialog box
A ------------------------------ xxxxxxxxxxxxxxxx
Trigger combinations -- A
Trigger qualifier ---------- x

Figure 22-4 7474 Edge-Triggered D Flip-Flop

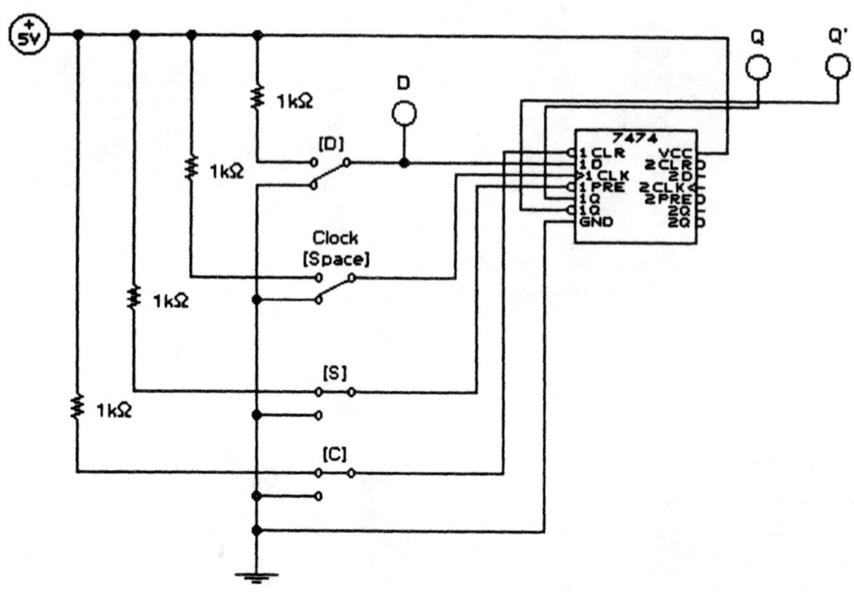

Figure 22-5a 7474 Edge-Triggered D Flip-Flop Timing (EWB Version 4)

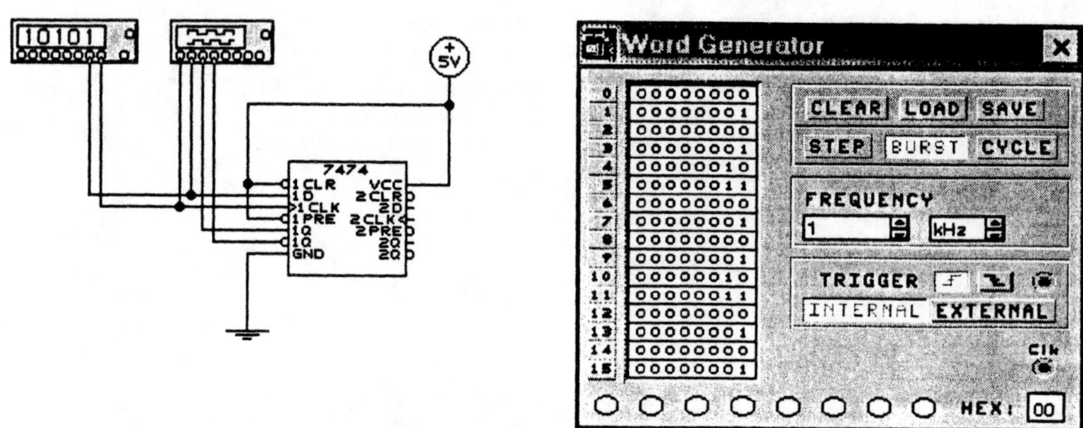

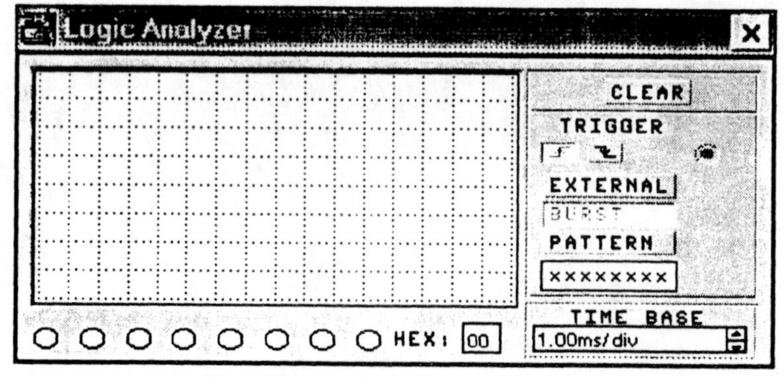

Figure 22-5b 7474 Edge-Triggered D Flip-Flop Timing (EWB Version 5)

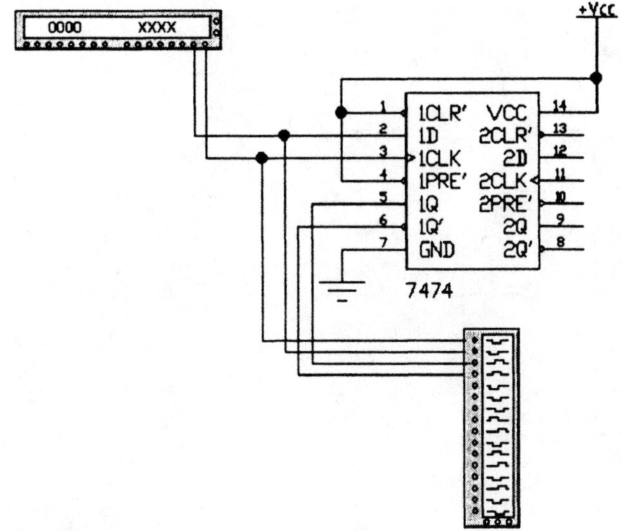

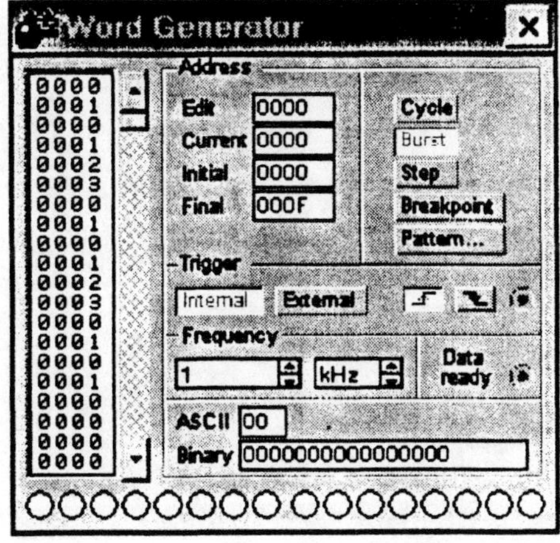

Logic Analyzer Settings
Clocks per division ---------------- 8

Clock Setup dialog box
Clock edge --------------- positive
Clock mode -------------- Internal
Internal clock rate ------ 10 kHz
Clock qualifier ----------- x
Pre-trigger samples --- 100
Post-trigger samples -- 1000
Threshold voltage (V) 2

Trigger Patterns dialog box
A ------------------------------ xxxxxxxxxxxxxxxx
Trigger combinations -- A
Trigger qualifier ---------- x

Figure 22-6 Edge-Triggered J-K Flip-Flop

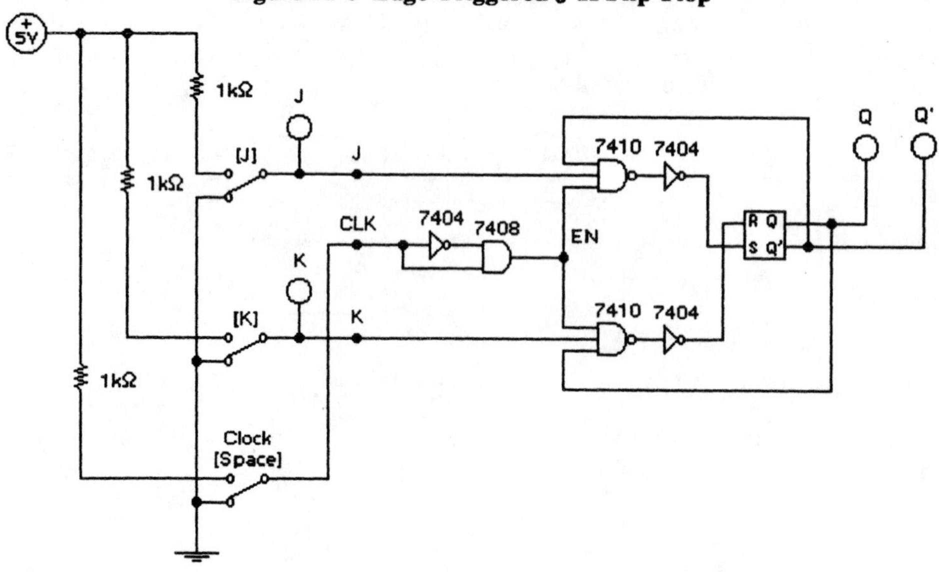

Figure 22-7 74112 Edge-Triggered J-K Flip-Flop

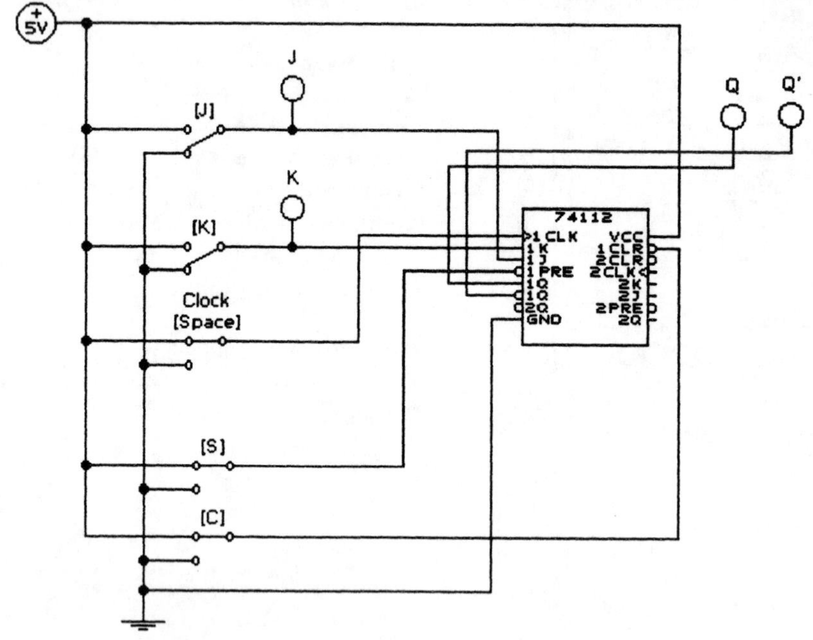

Figure 22-8a 74112 Edge-Triggered J-K Flip-Flop Timing (EWB Version 4)

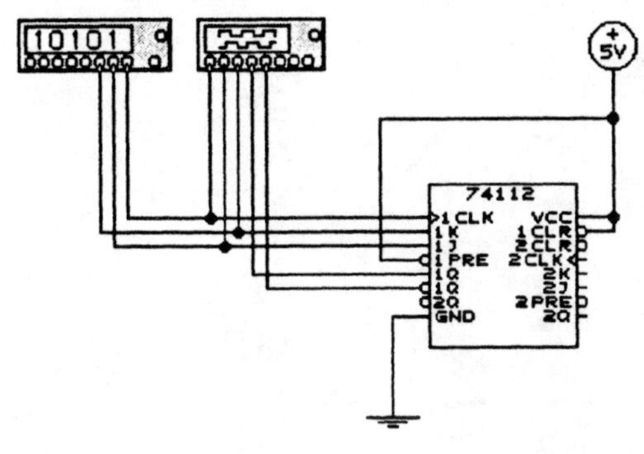

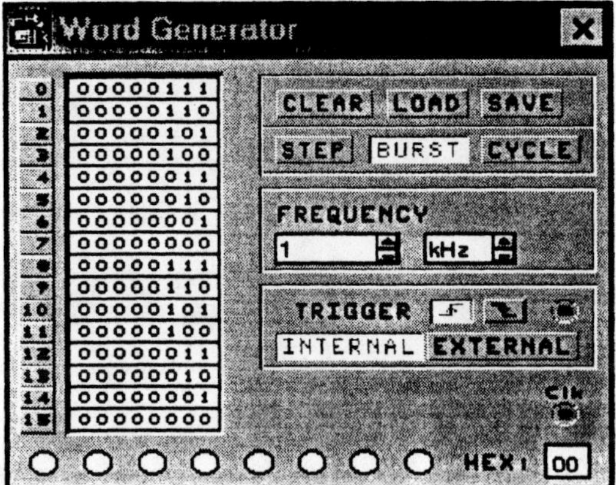

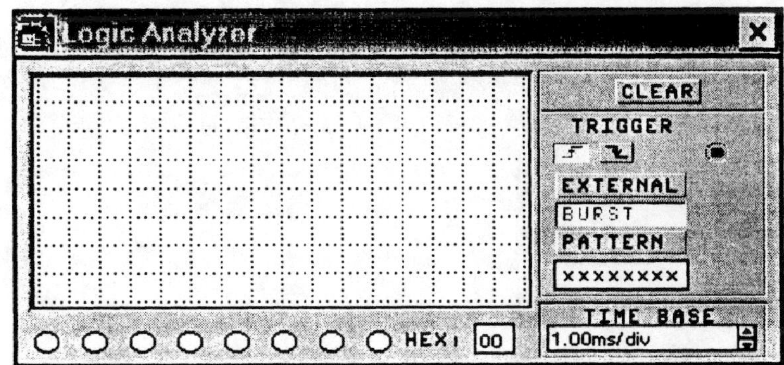

Figure 22-8b 74112 Edge-Triggered J-K Flip-Flop Timing (EWB Version 5)

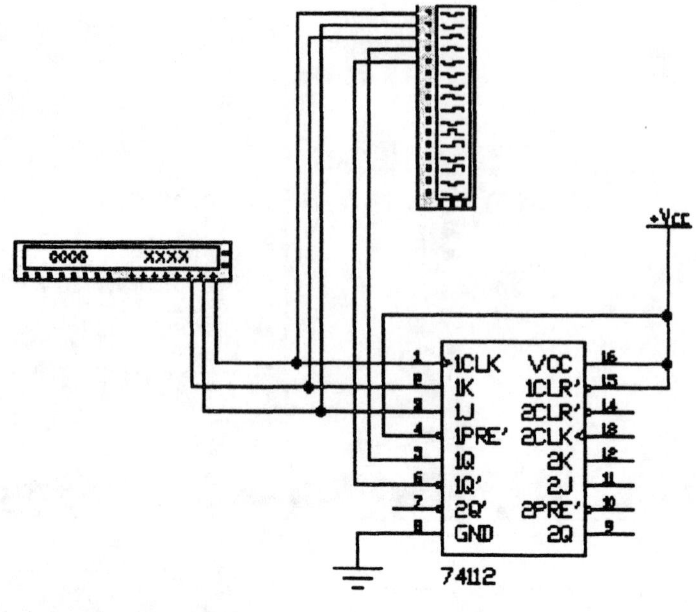

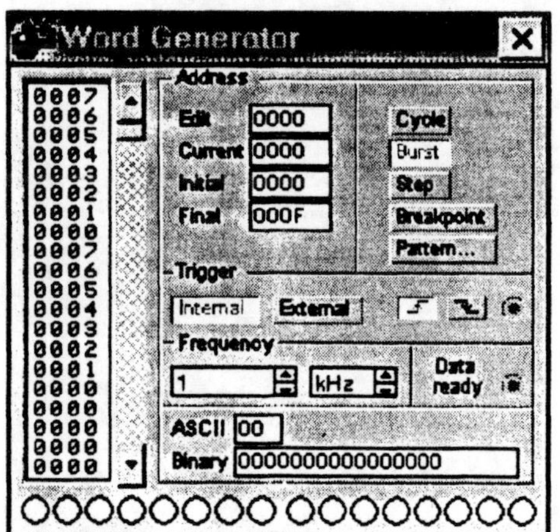

Logic Analyzer Settings
Clocks per division ---------------- 8

Clock Setup dialog box
Clock edge --------------- positive
Clock mode -------------- Internal
Internal clock rate ------ 10 kHz
Clock qualifier ----------- x
Pre-trigger samples --- 100
Post-trigger samples -- 1000
Threshold voltage (V) 2

Trigger Patterns dialog box
A ------------------------------ xxxxxxxxxxxxxxxx
Trigger combinations -- A
Trigger qualifier ---------- x

Procedure:

1. Pull down the File menu and open FIG22-1. You are looking at an S-R latch wired as a positive-edge-triggered S-R flip-flop. The INVERTER (IC1) and the AND gate (IC2A) form an edge detector circuit that provides a short duration pulse at the edge detector output (EN) on the positive edge of the clock pulse at the edge detector input (CLK). See the Preparation section for more details about the operation of this circuit.

NOTE: If this experiment is performed in a hardwired laboratory, the S-R latch should be replaced by the NOR gate S-R latch circuit in Figure 21-1, Experiment 21.

2. The S, R, and CLOCK switches should be down (0). Click the On-Off switch to run the analysis. If the output is clear (Q=0), press the S key on the computer keyboard to raise input S to binary one (1). If the output is set (Q=1), press the R key on the computer keyboard to raise input R to binary one (1).

Question: Did the output (Q) change state (set or clear)? Explain.

3. Press the space bar on the keyboard to make the CLK input rise from zero (0) to one (1) to produce a positive edge.

Question: Did the output (Q) change state (set or clear)? Explain.

4. Press the S and R keys to change inputs S and R so that they are the inverse of the previous values (S=0 and R=1, or S=1 and R=0). Press the space bar on the keyboard to make the CLK input drop from one (1) to zero (0) to produce a negative edge.

Questions: Did the output (Q) change state (set or clear)? Explain.

Is this a positive or a negative edge-triggered flip-flop?

5. Press the space bar on the keyboard again to make the CLK input rise from zero (0) to one (1) to produce a positive edge.

Question: Did the output (Q) change state (set or clear)? Explain.

6. Set both the S and R inputs to binary zero (0). Keep pressing the space bar on the keyboard to produce a series of positive clock edges.

Question: What happened to the flip-flop output (Q)? Explain.

7. Set both the S and R inputs to binary one (1). Keep pressing the space bar on the keyboard to produce a series of positive clock edges.

Question: What happened to the flip-flop output (Q)? Explain.

8. Click the On-Off switch to stop the analysis run. Pull down the File menu and open FIG22-2. You are looking at an S-R latch wired as a positive-edge-triggered D flip-flop. See the Preparation section for more details about the operation of this circuit.

NOTE: If this experiment is performed in a hardwired laboratory, the S-R latch should be replaced by the NOR gate S-R latch circuit in Figure 21-1, Experiment 21.

9. Both the D and the CLOCK switches should be down (0). Click the On-Off switch to run the analysis. If the output is clear (Q=0), press the D key on the computer keyboard to raise input D to binary one (1). If the output is set (Q=1), press the D key twice to raise input D to binary one (1) and back to binary zero (0).

Question: Did the output (Q) change state (set or clear)? Explain.

10. Press the space bar on the keyboard to make the CLK input rise from zero (0) to one (1) to produce a positive edge.

Question: Did the output (Q) change state (set or clear)? Explain.

11. Press the D key on the keyboard to make input D be the inverse of the previous value. Press the space bar on the keyboard to make the CLK input drop from one (1) to zero (0) to produce a negative edge.

Question: Did the output (Q) change state (set or clear)? Explain.

12. Press the space bar on the keyboard again to make the CLK input rise from zero (0) to one (1) to produce a positive edge.

Questions: Did the output (Q) change state (set or clear)? Explain.

Is this a positive or a negative edge-triggered flip-flop?

13. Click the On-Off switch to stop the analysis run. Pull down the File menu and open FIG22-3. You are looking at a circuit that will display the timing of a positive-edge-triggered D flip-flop. The word generator and logic analyzer settings should be as shown in Figure 22-3.

NOTE: If this experiment is being performed in a hardwired laboratory, use two pulse generators in place of the word generator and a dual trace oscilloscope in place of the logic analyzer if a logic analyzer is not available. Also, the S-R latch should be replaced by the NOR gate S-R latch circuit in Figure 21-1, Experiment 21.

14. Click the On-Off switch to run the analysis. The word generator is applying a series of clock pulses to the clock input (CLK) and a pulse waveshape to the D input of the flip-flop. The logic analyzer is monitoring the CLK input (green), the edge detector output EN (brown), the D input (red), output Q (blue), and inverted output Q'(black).

Questions: What do you notice about the output of the edge detector circuit (brown curve plot) compared to the CLK input (green curve plot)?

Does the flip-flop output Q (blue curve plot) only change state on a positive clock edge?

What determines whether the flip-flop output Q (blue curve plot) sets (Q=1) or clears (Q=0)?

What is the relationship between the flip-flop output Q (blue curve plot) and the inverted output Q' (black curve plot)?

15. Click the On-Off switch to stop the analysis run. Pull down the File menu and open FIG22-4. You will test a 7474 positive-edge-triggered D flip-flop with active low asynchronous preset (PRE) and clear (CLR) inputs. See the Preparation section for a detailed discussion of the 7474 flip-flop.

16. The D and CLOCK switches should be down (0). The S (set) and C (clear) switches should be up (1). Click the On-Off switch to run the analysis. Repeat Steps 9–12 for the circuit in Figure 22-4.

Question: Were the results the same as in Steps 9–12?

17. If the output is clear (Q=0), press the S key on the computer keyboard to drop the active low preset (PRE) input to binary zero (0). If the output is set (Q=1), press the C key to drop the active low clear (CLR) input to binary zero (0).

Question: Did the output (Q) change state without a positive clock edge being applied? Explain.

18. Press the S or C key on the keyboard so that both switches are up (1).

Question: Did the output (Q) change state? Explain.

19. Repeat Step 17.

Questions: Did the output (Q) change state without a positive clock edge being applied?

Are the 7474 preset (PRE) and clear (CLR) inputs asynchronous or synchronous inputs?

20. Click the On-Off switch to stop the analysis run. Pull down the File menu and open FIG22-5. You are looking at a circuit that will display the timing of the 7474 positive-edge-triggered D flip-flop. The word generator and logic analyzer settings should be as shown in Figure 22-5.

> NOTE: If this experiment is being performed in a hardwired laboratory, use two pulse generators in place of the word generator and a dual trace oscilloscope in place of the logic analyzer if a logic analyzer is not available.

21. Click the On-Off switch to run the analysis. The word generator is applying a series of clock pulses to the clock input (CLK) and a pulse waveshape to the D input of the flip-flop. The logic analyzer is monitoring the CLK input (green), the D input (red), output Q (blue), and inverted output Q'(black).

Questions: Does the flip-flop output Q (blue) change state on positive or negative clock edges?

What determines whether the flip-flop output Q (blue) sets (Q=1) or clears (Q=0)?

How did your results compare with the results for the D flip-flop in Figure 22-3 (Steps 13 and 14)?

22. Click the On-Off switch to stop the analysis run. Pull down the File menu and open FIG22-6. You are looking at an S-R latch wired as a positive-edge-triggered J-K flip-flop. See the Preparation section for more details about the operation of this circuit.

NOTE: If this experiment is performed in a hardwired laboratory, the S-R latch should be replaced by the NOR gate S-R latch circuit in Figure 21-1, Experiment 21.

23. The J,K, and the CLOCK switches should be down (0). Click the On-Off switch to run the analysis. If the output is clear (Q=0), press the J key on the computer keyboard to raise input J to binary one (1). If the output is set (Q=1), press the K key to raise input K to binary one (1). Press the space bar on the keyboard to make the CLK input rise from zero (0) to one (1) to produce a positive edge.

Question: Did the output (Q) change state (set or clear) on the positive edge of the clock?

24. Press the J and K keys to change inputs J and K so that they are the inverse of the previous values. Press the space bar on the keyboard to make the CLK input drop from one (1) to zero (0) to produce a negative edge.

Questions: Did the output (Q) change state (set or clear)?

Is this a positive or negative edge-triggered flip-flop?

25. Press the space bar on the keyboard to make the CLK input rise from zero (0) to one (1) to produce a positive edge.

Questions: Did the output (Q) change state (set or clear)?

Which is the set input and which is the clear input (J or K) on a J-K flip-flop?

26. Set both the J and K inputs to binary zero (0). Keep pressing the space bar on the keyboard to produce a series of positive clock edges.

Question: What happened to the flip-flop output (Q)? Explain.

27. Set both the J and K inputs to binary one (1). Keep pressing the space bar on the keyboard to produce a series of positive clock edges.

Questions: What happened to the flip-flop output (Q)? Explain.

What is the major difference between the J-K flip-flop and the S-R flip-flop?

28. Click the On-Off switch to stop the analysis run. Pull down the File menu and open FIG22-7. You will test a 74112 negative-edge-triggered J-K flip-flop with active low asynchronous preset (PRE) and clear (CLR) inputs. See the Preparation section for a detailed discussion of the 74112 J-K flip-flop.

29. The J and K switches should be down (0). The CLOCK, S (set) and C (clear) switches should be up (1). Click the On-Off switch to run the analysis. If the output is clear (Q=0), press the J key on the computer keyboard to raise input J to binary one (1). If the output is set (Q=1), press the K key to raise input K to binary one (1). Press the space bar on the keyboard to make the CLK input drop from one (1) to zero (0) to produce a negative edge.

Question: Did the output (Q) change state (set or clear) on the negative edge of the clock pulse?

30. Change inputs J and K so that they are the inverse of the previous values. Press the space bar on the keyboard to make the CLK input rise from zero (0) to one (1) to produce a positive edge.

Questions: Did the output (Q) change state (set or clear)?

Is this a positive or negative edge-triggered flip-flop?

31. Press the space bar on the keyboard to make the CLK input drop from one (1) to zero (0) to produce a negative edge.

Questions: Did the output (Q) change state (set or clear)?

Which is the set input and which is the clear input (J or K) on a J-K flip-flop?

32. Set both the J and K inputs to binary zero (0). Keep pressing the space bar on the keyboard to produce a series of negative clock edges.

Question: What happened to the flip-flop output (Q)? Explain.

33. Set both the J and K inputs to binary one (1). Keep pressing the space bar on the keyboard to produce a series of negative clock edges.

Questions: What happened to the flip-flop output (Q)?

How did your results for the 74112 J-K flip-flop compare with the results for the J-K flip-flop in Figure 22-6 (Steps 22-27)?

34. Keep pressing the space key on the keyboard until the output is clear (Q=0). Now press the S (set) key to drop the 74112 preset (PRE) input to binary zero (0).

Questions: What happened to the flip-flop output (Q)?.

Did the output change without a clock pulse being applied? Explain.

35. Press the S (set) key to raise the 74112 preset (PRE) input to binary one (1). Press the C (clear) key to drop the 74112 clear (CLR) input to binary zero (0).

Questions: What happened to the flip-flop output (Q)? Explain.

Did the output change without a clock pulse being applied?

Are the 74112 preset (PRE) and clear (CLR) inputs asynchronous or synchronous inputs? Explain.

36. Click the On-Off switch to stop the analysis run. Pull down the File menu and open FIG22-8. You are looking at a circuit that will display the timing of the 74112 negative-edge-triggered J-K flip-flop. The word generator and logic analyzer settings should be as shown in Figure 22-8.

NOTE: If this experiment is being performed in a hardwired laboratory, use three pulse generators in place of the word generator and a dual trace oscilloscope in place of the logic analyzer if a logic analyzer is not available.

37. Click the On-Off switch to run the analysis. The word generator is applying a series of clock pulses to the clock input (CLK), a pulse waveshape to the J input, and a different pulse waveshape to the K input of the flip-flop. The logic analyzer is monitoring the CLK input (green), the J input (red), the K input (brown), output Q (blue), and inverted output Q'(black).

Questions: Does the flip-flop output Q (blue) change state on the positive or negative clock edges?

Does the flip-flop output Q (blue) toggle when the J and K inputs are both binary one (1) during a negative clock edge?

Does the flip-flop output Q (blue) set (Q=1) when J=1 and K=0 during a negative clock edge?

Does the flip-flop output Q (blue) clear (Q=0) when J=0 and K=1 during a negative clock edge?

Does the flip-flop output Q (blue) stay the same when J=0 and K=0 during a negative clock edge?

Name ______________________

Date ______________________

EXPERIMENT

23 Monostable and Astable Multivibrators

Objectives:

1. Investigate the operation of a monostable multivibrator (one-shot).
2. Investigate the operation of a 555 timer wired as an astable multivibrator.

Materials:

One 5 V dc power supply
One logic switch
One logic probe light
One monostable multivibrator (one-shot) (1-74121 IC)
One timer (1-555 IC)
One square wave generator
One dual-trace oscilloscope
One 1 μF capacitor
One 100 μF capacitor
Two 0.01 μF capacitors
One 0.02 μF capacitor
Resistors—200 kΩ, 100 kΩ, 72 kΩ, 48 kΩ, 10 kΩ, 5 kΩ, 1 kΩ

Preparation:

Latches and flip-flops, which were studied in the previous experiments, are often referred to as bistable multivibrators because the output has two stable states (high or low). A monostable multivibrator, otherwise known as a one-shot, has only one stable output state. When it is triggered, the output will go to the unstable state (Q=1) for a predetermined period of time, and then return to the stable state (Q=0). The time period that the output will stay in the unstable state (Q=1) is determined by the RC circuit connected to the one-shot. A nonretriggerable one-shot will not respond to any additional trigger pulses during the time that it is in the unstable state (Q=1). Therefore, the time that the output is unstable (Q=1) for a nonretriggerable one-shot will always be dependent on the RC time constant only. A retriggerable one-shot will respond to trigger pulses during the time that the output is in the unstable state (Q=1). Therefore, the unstable state (Q=1) for a retriggerable one-shot will restart on each trigger pulse, causing the time that the output is unstable (Q=1) to depend on the trigger pulse timing.

An astable multivibrator doesn't have any stable output state. Therefore, it continuously switches between two unstable states without any triggering. This results in a square wave output that is useful for providing clock pulses for synchronous digital circuits. An astable multivibrator is often referred to as a free-running pulse oscillator or generator.

The circuit in Figure 23-1 will be used to demonstrate the operation of a monostable multivibrator (one-shot). The CLOCK switch will be used to provide a negative clock edge to the negative-edge-triggered trigger input (A2). The logic probe light will monitor the one-shot output (Q). The time constant of the RC circuit connected to the one-shot will determine the time period (t_w) that the one-shot output will stay in the unstable state (Q=1). This time period (t_w) can be calculated from the equation

$$t_w = 0.7(R)(C)$$

The circuit in Figure 23-2 will be used to demonstrate the timing of a monostable multivibrator (one-shot). The pulse generator will apply a series of short duration trigger pulses to the one-shot negative-edge-triggered trigger input (A2). The oscilloscope will monitor the one-shot trigger input (A2) and the output (Q).

The 555 timer in Figure 23-3 is wired as an astable multivibrator (free-running pulse generator). The external components RA, RB, and C determine the frequency of oscillation (f) and the duty cycle. Capacitor C2 is strictly for decoupling the control input (CON) from noise pickup and has no effect on the oscillating frequency. In some cases it can be left out. The frequency of oscillation (f) is the inverse of the time period for one pulse cycle (T). Therefore,

$$f = \frac{1}{T}$$

The duty cycle is the time that the output is high (t_2) divided by the total time period for one pulse cycle (T), expressed as a percentage. Therefore,

$$\text{Duty cycle} = \frac{t_2}{T} \times 100\%$$

The expected frequency of oscillation (f) and the expected duty cycle can be calculated from the values of RA, RB, and C by first calculating the expected time that the output is low (t_1), the expected time that the output is high (t_2), and the expected time period for one pulse cycle (T) from the equations

$$t_1 = 0.7(RB)(C)$$
$$t_2 = 0.7(RA + RB)(C)$$
$$T = t_1 + t_2$$

For obtaining a 50% duty cycle,

$$\frac{t_2}{T} = 0.5$$

Therefore,

$$t_2 = 0.5(T) = \frac{T}{2}$$

This can only be achieved if RB is much larger than RA so that

$$t_2 = 0.7(RA + RB)(C) \approx 0.7(RB)(C) = t_1$$

Therefore,

$$T = t_1 + t_2 \approx t_2 + t_2 = 2(t_2)$$

and $t_2 \approx \frac{T}{2}$

Figure 23-1 Monostable Multivibrator (One-Shot)

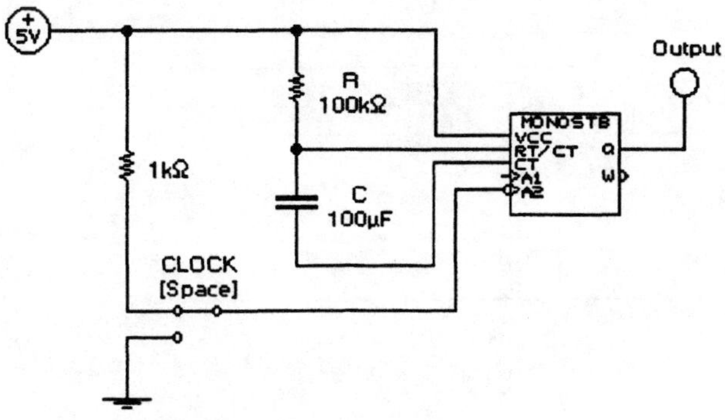

Figure 23-2 Monostable Multivibrator (One-Shot) Timing

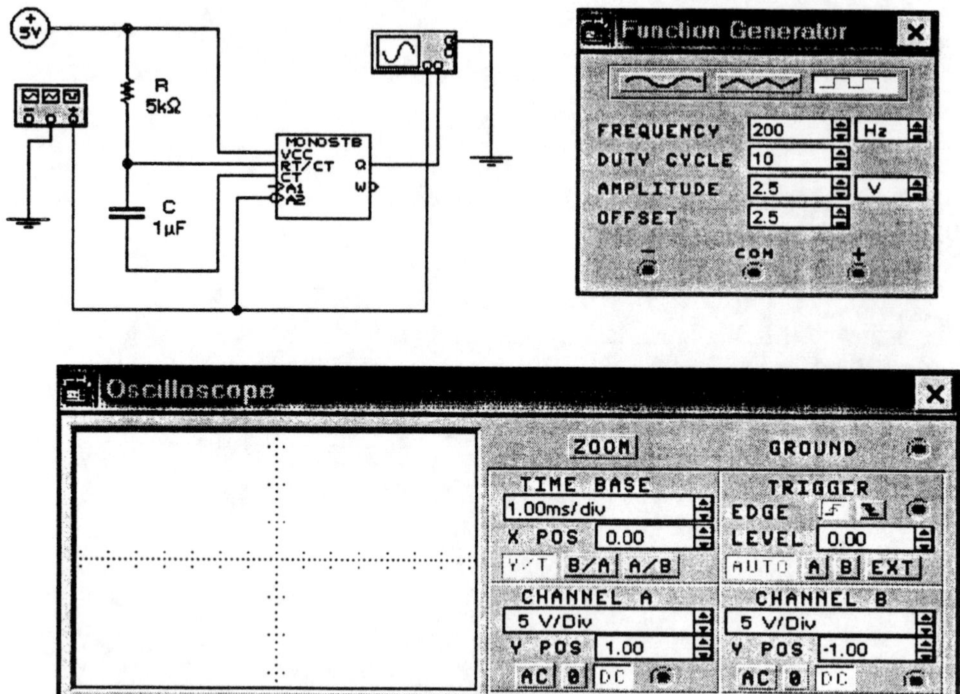

Figure 23-3 555 Astable Multivibrator

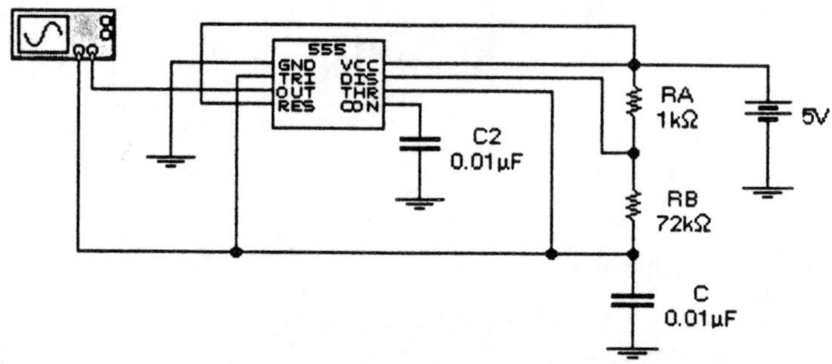

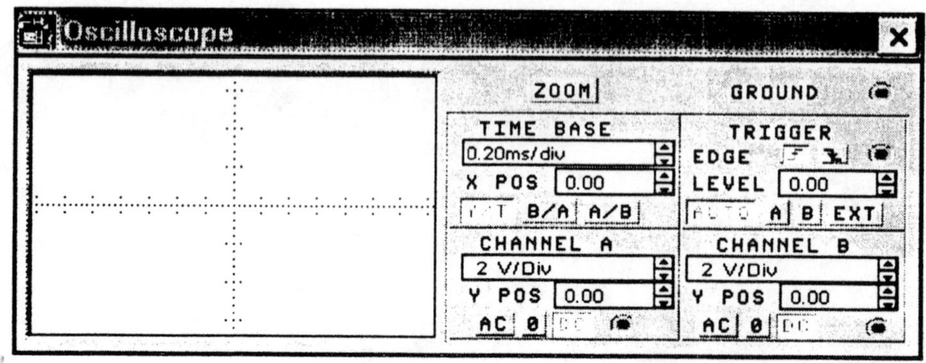

Procedure:

1. Pull down the File menu and open FIG23-1. You are looking at a circuit that will demonstrate the operation of a monostable multivibrator (one-shot). The logic probe light will monitor the one-shot output (Q). The CLOCK logic switch will provide a negative clock edge on the one-shot negative-edge-triggered trigger input (A2). The CLOCK switch should be up (1). Click the On-Off switch to run the analysis. Press the space bar on the computer keyboard to apply a negative clock edge to the one-shot trigger input (A2) and measure and record the time (t_w) that the one-shot output is high (Q=1).

   t_w = ______________

> NOTE: If this experiment is performed in a hardwired laboratory, use a 74121 monostable multivibrator (one-shot) and connect trigger input B to +5 V (1) and trigger input A1 to ground (0).

2. Based on the value of resistor R and capacitor C, calculate the expected time (t_w) that the output will be high (Q=1).

Question: How did your calculated value for t_w compare with the measured value in Step 1?

3. Click the On-Off switch to stop the analysis run. Change resistor R to 200 kΩ. Make sure that the CLOCK switch is up (1). Click the On-Off switch to run the analysis again. Press the space bar on the computer keyboard to apply a negative clock edge to the one-shot trigger input (A2) and measure and record the new time (t_w) that the one-shot output is high (Q=1).

   t_w = ______________

Question: How did the new value for t_w compare with the measured value in Step 1? Explain the reason for any difference.

4. Click the On-Off switch to stop the analysis run. Pull down the File menu and open FIG23-2. You are looking at a monostable multivibrator (one-shot) connected to a pulse generator and oscilloscope. The pulse generator and oscilloscope settings should be as shown in Figure 23-2.

> NOTE: If this experiment is performed in a hardwired laboratory, use a 74121 monostable multivibrator (one-shot) and connect trigger input B to +5 V (1) and trigger input A1 to ground (0).

5. Click the On-Off switch to run the analysis. The pulse generator is applying a series of short duration trigger pulses to the one-shot negative-edge-triggered trigger input (A2). The oscilloscope is monitoring the one-shot trigger input, A2 (red curve plot), and the one-shot output, Q (blue curve plot). Measure and record the high level pulse width (t_w) at the one-shot output (blue curve plot).

   t_w = ______________

Question: Is the one-shot being triggered on the positive edge or the negative edge of the trigger input pulse (red curve plot)? Explain how you determined the answer.

6. Based on the value of resistor R and capacitor C, calculate the expected pulse width (t_w) at the output (Q) of the one-shot.

Question: How did your calculated value for the output pulse width (t_w) compare with the measured value in Step 5?

7. Click the On-Off switch to stop the analysis run. Change the value of resistor R to 10 kΩ. Click the On-Off switch to run the analysis again. Measure and record the new pulse width (t_w) at the one-shot output (blue curve plot).

   t_w = ______________

8. With the new value of R, calculate the new expected pulse width (t_w).

Questions: How did your new calculated value for the output pulse width (t_w) compare with the measured value in Step 7?

Based on the curve plots for the one-shot output (blue) and the trigger input (red), is this a nonretriggerable or retriggerable monostable multivibrator? Explain how you determined the answer.

9. Click the On-Off switch to stop the analysis run. Pull down the File menu and open FIG23-3. You are looking at a 555 timer wired as an astable multivibrator (free-running pulse generator). The oscilloscope settings should be as shown in Figure 23-3. Click the On-Off switch to run the analysis. After the simulation reaches steady-state (approximately 20 msec. elapsed simulation time), press the F9 key on the computer keyboard to "pause" the simulation.

10. Measure and record the time that the output is low (t_1), the time that the output is high (t_2), and the total time period for one cycle (T) of the output (blue curve plot).

t_1 = ___________ t_2 = ___________ T = ___________

11. Measure and record the trigger voltage (red) that switches the output high (V_{high}) and the trigger voltage (red) that switches the output low (V_{low}).

V_{high} = ___________ V_{low} = ___________

Questions: What percentage of Vcc is V_{high}?

What percentage of Vcc is V_{low}?

12. Based on the time period (T) measured in Step 10, calculate the pulse frequency (f) in hertz.

13. Based on t_1, t_2, and T measured in Step 10, calculate the duty cycle.

14. Based on the values of RA, RB, and C for the circuit in Figure 23-3, calculate the expected values of t_1, t_2, and T.

Question: How did your new calculated values for t_1, t_2, and T compare with the measured values in Step 10?

15. Click the On-Off switch to stop the analysis run. Change resistor RA and resistor RB to 48 kΩ. Click the On-Off switch to run the analysis. After the simulation reaches steady-state (approximately 20 msec. elapsed simulation time), press the F9 key on the computer keyboard to "pause" the simulation.

16. Measure and record the time that the output is low (t_1), the time that the output is high (t_2), and the total time period for one cycle (T) of the output (blue curve plot).

t_1 = ____________ t_2 = ____________ T = ____________

17. Based on the time period (T) measured in Step 16, calculate the new pulse frequency (f) in hertz.

18. Based on t_1, t_2, and T measured in Step 16, calculate the new duty cycle.

Question: How does the new duty cycle compare with the duty cycle calculated in Step 13? Explain the reason for any difference.

19. Based on the new values of RA and RB, and the value of C, for the circuit in Figure 23-3, calculate the new expected values of t_1, t_2, and T.

Questions: How did your new calculated values for t_1, t_2, and T compare with the measured values in Step 16?

How did the new values of t_1, t_2, and T compare with the values in Steps 10 and 14? Explain the reason for any difference.

20. Click the On-Off switch to stop the analysis run. Change capacitor C to 0.02 μF. Click the On-Off switch to run the analysis. After the simulation reaches steady-state (approximately 20 msec. elapsed simulation time), press the F9 key on the computer keyboard to "pause" the simulation.

21. Measure and record the time that the output is low (t_1), the time that the output is high (t_2), and the total time period for one cycle (T) of the output (blue curve plot).

t_1 = __________ t_2 = __________ T = __________

Question: How did the values for t_1, t_2, and T compare with the measured values in Step 16 with C = 0.01 μF? Explain the reason for any difference.

22. Based on the values of t_1, t_2, and T, calculate the new duty cycle.

Question: How does the new duty cycle compare with the duty cycle calculated in Step 18 with C = 0.01 μF? Explain your answer.

23. Based on the new time period (T), calculate the new pulse frequency (f) in hertz.

Name ____________________

Date ____________________

Registers and Data Storage

Objectives:

1. Demonstrate the application of flip-flops in serial and parallel registers.
2. Demonstrate parallel register data transfer.
3. Demonstrate serial register data transfer.
4. Investigate the operation of a 74173 four-bit parallel in/parallel out register.
5. Investigate the operation of a 74194 universal four-bit bidirectional shift register.
6. Observe shift register timing.

Materials:

One 5 V dc power supply
Nine logic switches
Eight logic probe lights
Four positive-edge-triggered D-type flip-flops (2-7474 ICs)
Seven two-input AND gates (2-7408 ICs)
Three two-input OR gates (1-7432 IC)
One INVERTER (1-7404 IC)
One four-bit parallel in/parallel out register (1-74173 IC)
One universal four-bit bidirectional shift register (1-74194 IC)
Two pulse generators
One logic analyzer or dual-trace oscilloscope

Preparation:

A register consists of a series of flip-flops for the purpose of storing and transferring binary data in a digital system. Each register flip-flop stores one bit of binary data. The storage capacity of the register is determined by the number of flip-flops, which determines the number of binary bits it can store. Registers can be loaded in serial one bit at a time, or in parallel with all bits loaded at one time. Binary data can be output from a register in serial one bit at a time, or in parallel with all bits output at one time. Register input and output is controlled by the positive or negative edge of a clock pulse applied to the register clock terminal. Some registers can also be input enabled or disabled and output enabled or disabled.

The circuit in Figure 24-1 consists of four positive-edge-triggered D-type flip-flops wired as a parallel in/parallel out register. Each flip-flop stores one bit of binary data. Because all four flip-flop clock inputs are connected together, they receive the positive clock edge at the same time. This causes the bits to be stored simultaneously in each flip-flop, making this a parallel in register. Because all four flip-flop outputs are being monitored simultaneously, this is also a parallel out register. This register can be cleared by clearing each flip-flop (Q=0). This is accomplished by connecting the active low asynchronous clear inputs of the flip-flops together to form an active low CLR' terminal. When a binary zero (0) is applied to the CLR' terminal, the register will clear (Q=0). The CLR' terminal must be brought back to binary one (1) if you want to store new binary data in the register, otherwise the register will stay cleared because the flip-flop asynchronous clear inputs override the clock inputs.

The circuit in Figure 24-2 consists of four positive-edge-triggered D-type flip-flops wired as a serial in/parallel out register. Each flip-flop stores one bit of binary data. Because all four flip-flop clock inputs are connected together, they receive the positive clock edge at the same time. Data is input to this register into the first flip-flop one bit at a time. Because each flip-flop output is connected to the next flip-flop input, each data bit is shifted right on each clock pulse positive edge, making this a serial in shift register. Therefore, it takes four clock pulses to store 4 bits of binary data in this register. Because all four flip-flop outputs are being monitored simultaneously, this is a parallel out register.

The circuit in Figure 24-3 consists of four positive-edge-triggered D-type flip-flops wired as a serial in/serial out register. Each flip-flop stores one bit of binary data. Because all four flip-flop clock inputs are connected together, they receive the positive clock edge at the same time. Data is input to this register into the first flip-flop one bit at a time. Because each flip-flop output is connected to the next flip-flop input, each data bit is shifted right on each clock pulse positive edge, making this a serial in shift register. Therefore, it takes four clock pulses to store 4 bits of binary data in this register. Because only the last flip-flop output is being monitored, this register must be clocked three times to shift all four bits to the output (Q). Therefore, this is a serial out register. Remember, the last flip-flop output can be read before any clock pulses are applied. This is the reason that only three clock pulses are needed to read the output data.

The circuit in Figure 24-4 consists of four positive-edge-triggered D-type flip-flops and some additional logic circuitry wired as a parallel in/serial out register. The S/L' input is the shift/load input. A binary zero (0) on this input will cause the logic circuitry to connect each input bit (D3–D0) to the D input on each flip-flop. Therefore, this register will parallel load the input data (D3–D0) on the positive edge of the clock pulse when the S/L' input is at binary zero (0), making this a parallel in register. A binary one (1) on the S/L' input will cause the logic circuitry to shift each stored bit to the next flip-flop on each positive edge of the clock pulse. Because only the last flip-flop is being monitored, a binary one (1) on the S/L' input makes it possible to read the register output (Q) one bit at a time, making this a serial out register.

The 74173 in Figure 24-5 is a 4-bit parallel in/parallel out register. This register has two active low input enables (G1 and G2) and two active low output enables (M and N). It also has an active high clear (CLR) and a positive-edge-triggered clock input (CLK). The 74173 internal circuity is similar to the circuitry in Figure 24-1 with some additional logic circuitry to control the input enable and output enable.

The 74194 in Figure 24-6 is a universal 4-bit bidirectional shift register. This register has a parallel load data input (D–A) and two serial load data inputs (SL and SR). Serial load data input SL is a shift left serial input, and serial load data input SR is a shift right serial input. Shift left means shifting from QD to QA (Q3 to Q0) and shift right means shifting from QA to QD (Q0 to Q3). The mode control inputs (S1 S0) determine whether the register shifts left, shifts right, parallel loads, or stays the same when a positive clock edge is applied to the CLK input. See Table 24-1 for the codes.

Table 24-1 74194 Mode Control Inputs

S1 S0	Result on CLK Edge
0 0	No change
0 1	Shift right (QA to QD)
1 0	Shift left (QD to QA)
1 1	Parallel load

The circuit in Figure 24-7 will display the timing of a 74194 shift register. The mode control input (S1 S0) determines the register mode. (See Table 24-1). The word generator will apply a series of clock pulses to the register clock input (CLK) and a pulse waveform to the serial data inputs (SL and SR). The logic analyzer will monitor the clock input (CLK), the serial data inputs (SL and SR), and the four outputs (QD–QA).

Figure 24-1 Parallel In/Parallel Out Register

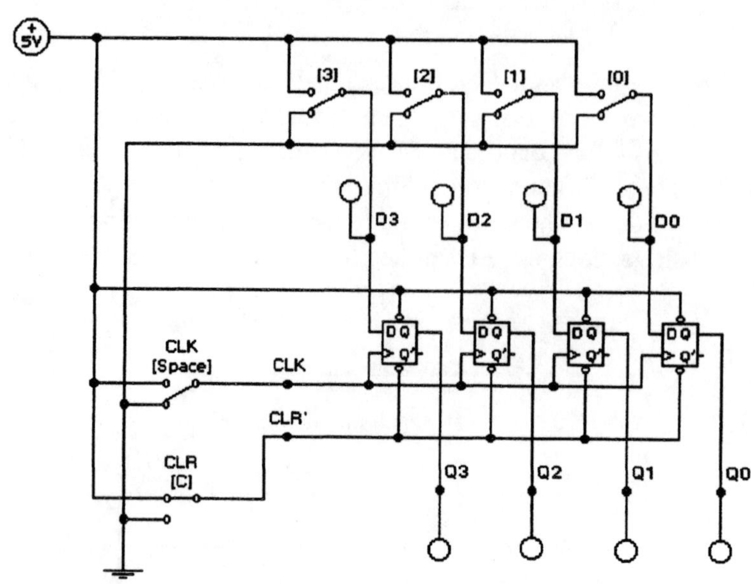

Figure 24-2 Serial In/Parallel Out Register

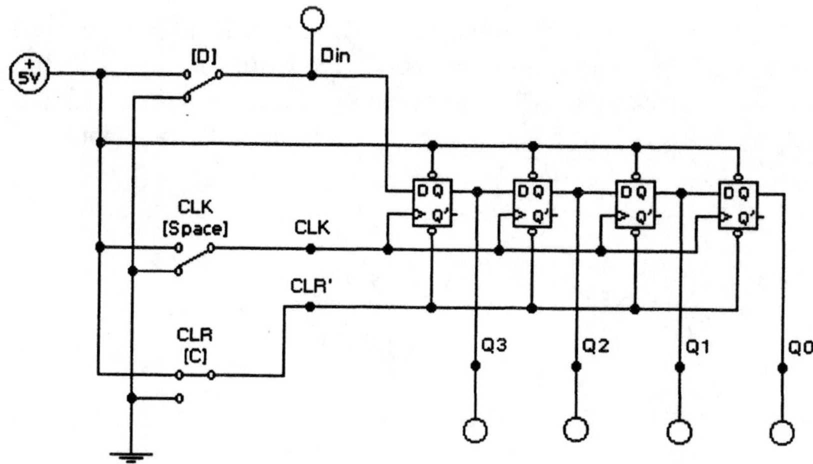

Figure 24-3 Serial In/Serial Out Register

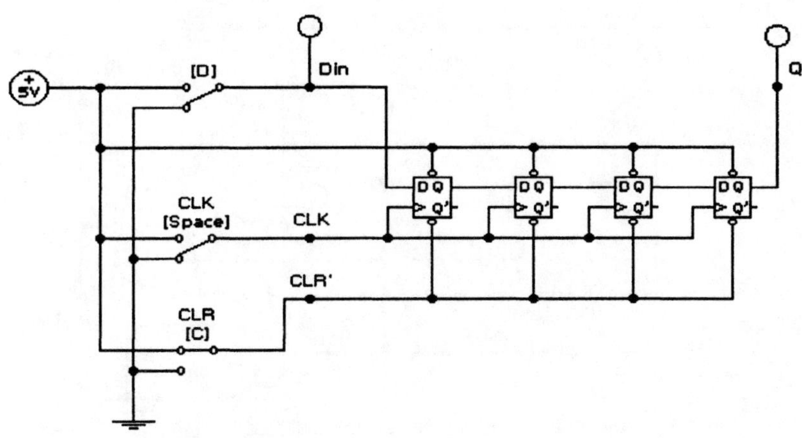

Figure 24-4 Parallel In/Serial Out Register

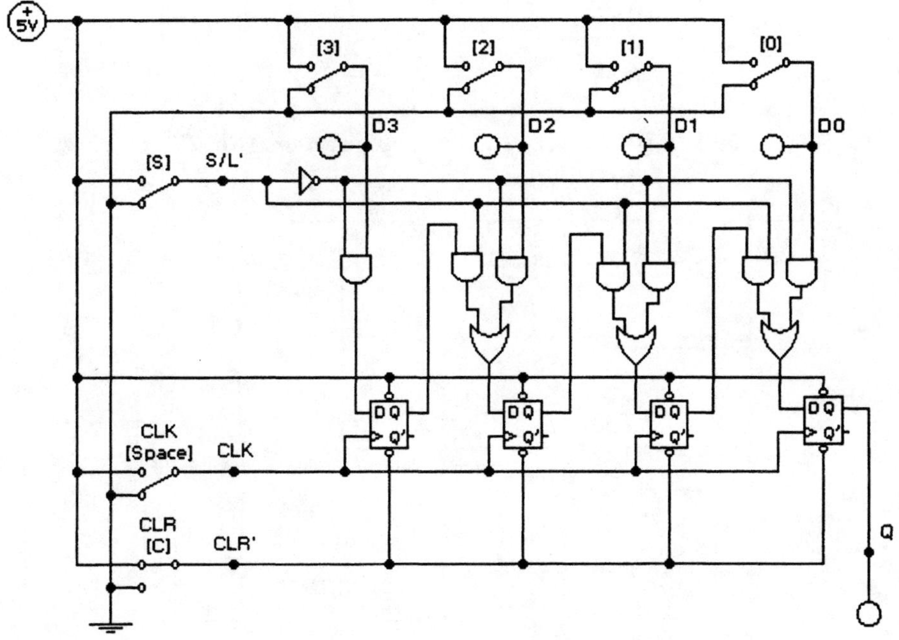

Figure 24-5 74173 Four-bit Parallel In/Parallel Out Register

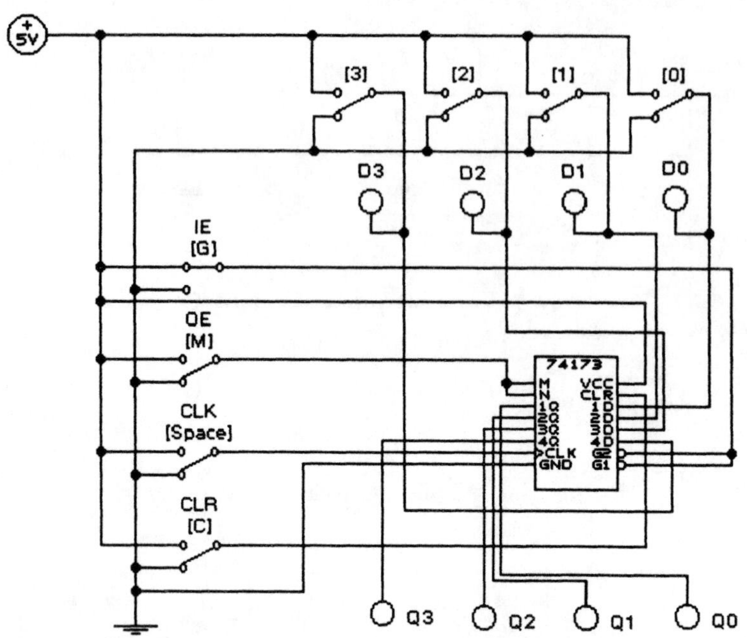

Figure 24-6 74194 Universal Four-bit Bidirectional Shift Register

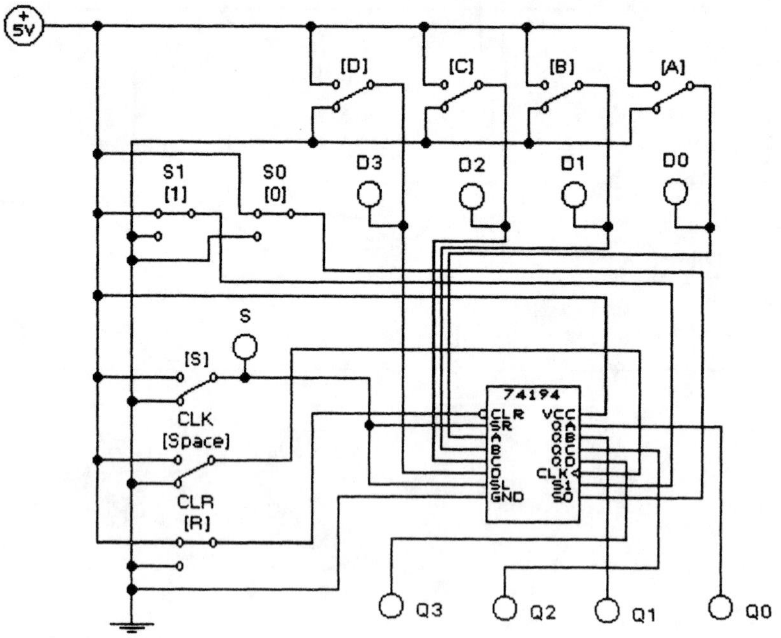

Figure 24-7a 74194 Shift Register Timing (EWB Version 4)

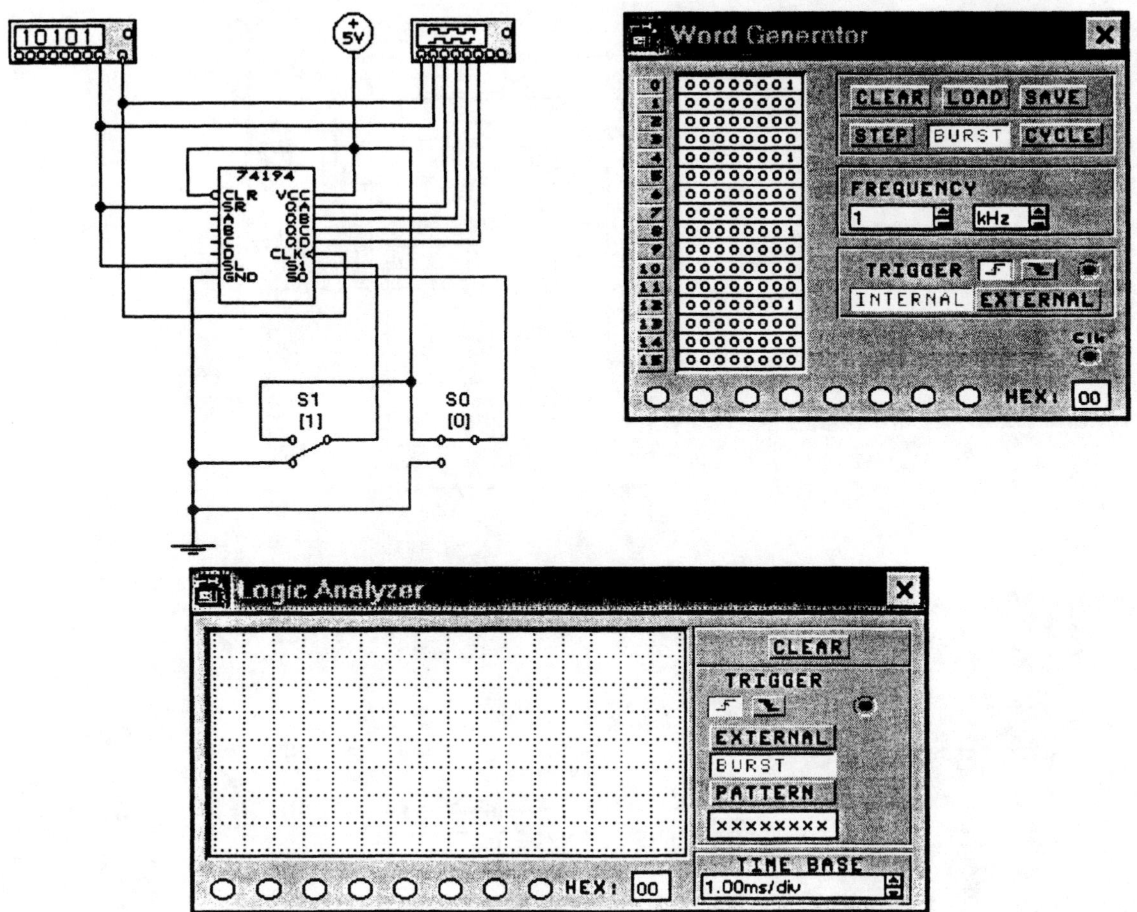

Figure 24-7b 74194 Shift Register Timing (EWB Version 5)

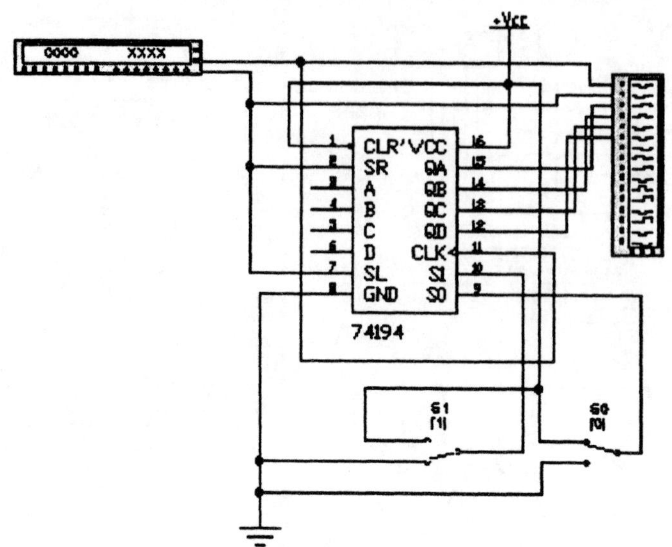

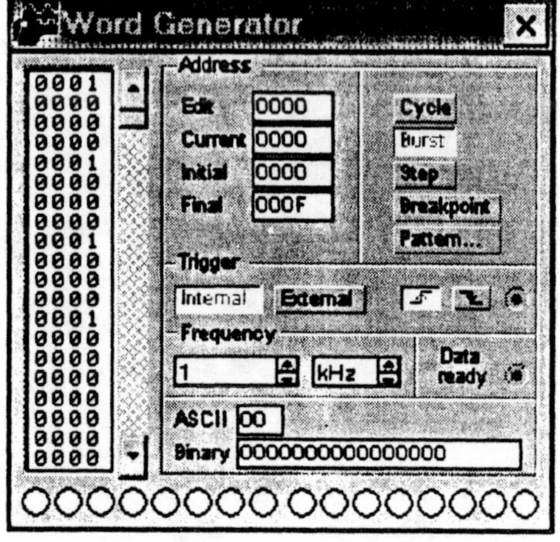

Logic Analyzer Settings
Clocks per division ---------------8

Clock Setup dialog box
Clock edge -------------- positive
Clock mode ------------- Internal
Internal clock rate ------ 10 kHz
Clock qualifier ---------- x
Pre-trigger samples --- 100
Post-trigger samples -- 1000
Threshold voltage (V) 2

Trigger Patterns dialog box
A ---------------------------- xxxxxxxxxxxxxxxx
Trigger combinations -- A
Trigger qualifier --------- x

Procedure:

1. Pull down the File menu and open FIG24-1. You are looking at four positive-edge-triggered D-type flip-flops wired as a parallel in/parallel out register. The CLR switch should be up (1) and the CLK switch should be down (0). Click the On-Off switch to run the analysis and press the C key on the computer keyboard to clear the register. Press the C key again to raise the CLR switch back to binary one (1).

> NOTE: If this experiment is being performed in a hardwired laboratory, use two 7474 ICs for the four positive-edge-triggered D-type flip-flops.

Questions: Is the clear input (CLR') active low or active high?

Why does the CLR switch need to be up (1) to store data in the register?

2. Set the data input switches (0–3) to any 4-bit binary number other than zero.

Question: Did the register output (Q3–Q0) change? Explain.

3. Press the space bar on the computer keyboard to make the CLK input rise from zero (0) to one (1) to produce a positive clock edge.

Questions: What did you observe at the register output (Q3–Q0)? Explain.

Is this register positive-edge-triggered or negative-edge-triggered?

Did the input data load in parallel or serial?

Is the data being read at the output in parallel or serial?

How many clock pulses were required to store the 4-bit binary input data?

4. Change the binary data input (D3–D0) by changing the data switches (0–3). Press the space bar on the computer keyboard to make the CLK input drop from one (1) to zero (0) to produce a negative clock edge.

Question: Did the register output (Q3–Q0) change? Explain.

5. Press the space bar on the computer keyboard again to make the CLK input rise from zero (0) to one (1) to produce a positive clock edge.

Question: What did you observe at the register output (Q3–Q0)? Explain.

6. Click the On-Off switch to stop the analysis run. Pull down the File menu and open FIG24-2. You are looking at four positive-edge-triggered D-type flip-flops wired as a serial in/parallel out register. The CLR switch should be up (1) and the CLK switch should be down (0). Click the On-Off switch to run the analysis and press the C key on the computer keyboard to clear the register. Press the C key again to raise the CLR switch back to binary one (1).

> NOTE: If this experiment is being performed in a hardwired laboratory, use two 7474 ICs for the four positive-edge-triggered D-type flip-flops.

7. Press the D key on the keyboard to set the data input switch (D) to binary one (up) so that Din=1.

Question: Did the register output (Q3–Q0) change? Explain.

8. Press the space bar on the computer keyboard to make the CLK input rise from zero (0) to one (1) to produce a positive clock edge.

Question: What did you observe at the register output (Q3–Q0)? Explain.

9. Press the D key on the keyboard again to set the data input switch (D) to binary zero (down) so that Din=0. Press the space bar on the computer keyboard to make the CLK input fall from one (1) to zero (0) to produce a negative clock edge, and press the space bar again to produce a positive clock edge.

Question: What did you observe at the register output (Q3–Q0)? Explain.

10. Repeat Steps 7–9.

Questions: What did you observe at the register output (Q3–Q0)?

Is this register positive-edge-triggered or negative-edge-triggered?

Did the input data load in parallel or serial?

Is the data being read at the output in parallel or serial?

How many clock pulses were required to store the 4-bit binary input data?

11. Click the On-Off switch to stop the analysis run. Pull down the File menu and open FIG24-3. You are looking at four positive-edge-triggered D-type flip-flops wired as a serial in/serial out register. The CLR switch should be up (1) and the CLK switch should be down (0). Click the On-Off switch to run the analysis and press the C key on the computer keyboard to clear the register. Press the C key again to raise the CLR switch back to binary one (1).

NOTE: If this experiment is being performed in a hardwired laboratory, use two 7474 ICs for the four positive-edge-triggered D-type flip-flops.

12. Press the D key on the keyboard to set the data input switch (D) to binary one (up) so that Din=1. Press the space bar on the computer keyboard to make the CLK input rise from zero (0) to one (1) to produce a positive clock edge.

13. Press the D key on the keyboard to set the data input switch (D) to binary zero (down) so that Din=0. Press the space bar on the computer keyboard to make the CLK input fall from one (1) to zero (0) to produce a negative clock edge and press the space bar again to produce a positive clock edge.

14. Repeat Steps 12 and 13.

Question: What did you observe at the register output (Q) after the fourth positive clock edge? Explain.

15. Press the space bar on the computer keyboard enough times to produce three positive clock edges and observe the register output (Q).

Questions: What did you observe at the register output (Q) on each positive edge of the clock?

Is this register positive-edge-triggered or negative-edge-triggered?

Did the input data load in parallel or serial?

Is the data being read at the output in parallel or serial?

How many clock pulses were required to store the 4-bit binary input data?

How many clock pulses were required to read the 4-bit binary output data?

16. Click the On-Off switch to stop the analysis run. Pull down the File menu and open FIG24-4. You are looking at four positive-edge-triggered D-type flip-flops and some additional logic circuitry wired as a parallel in/serial out register. The S/L' input is the shift/load input. A binary zero (0) on this input will cause this register to parallel load the input data (D3–D0) on the positive edge of the clock pulse. A binary one (1) on this input will cause the register to shift each stored bit to the output (Q) one bit at a time on each positive edge of the clock pulse. The CLR switch should be up (1) and the CLK switch and the S switch should be down (0).

NOTE: If this experiment is being performed in a hardwired laboratory, use two 7474 ICs for the positive-edge-triggered D-type flip-flops, two 7408 ICs for the 2-input AND gates, one 7432 IC for the 2-input OR gates, and one 7404 IC for the INVERTER.

17. Click the On-Off switch to run the analysis and press the C key on the computer keyboard to clear the register. Press the C key again to raise the CLR switch back to binary one (1).

18. Set the data input switches (0–3) to any 4-bit binary number other than zero. Make sure the S switch is down (0). Press the space bar on the computer keyboard to make the CLK input rise from zero (0) to one (1) to produce a positive clock edge. The binary input data (D3–D0) should have been parallel loaded into the register on the positive edge of the clock pulse even though all of the outputs are not being monitored.

19. Press the S key on the computer keyboard to set the S/L' input to binary one (1). Press the space bar on the computer keyboard enough times to produce three positive clock edges and observe the register output (Q).

Questions: What did you observe at the register output (Q) on each positive edge of the clock pulse?

Is this register positive-edge-triggered or negative-edge-triggered?

Did the input data load in parallel or serial?

Is the data being read at the output in parallel or serial?

How many clock pulses were required to store the 4-bit binary input data?

How many clock pulses were required to read the 4-bit binary output data?

20. Click the On-Off switch to stop the analysis run. Pull down the File menu and open FIG24-5. You will test a 74173 4-bit parallel in/parallel out register. This register has two active low input enables (G1 and G2) which are activated by switch IE, and two active low output enables (M and N), which are activated by switch OE. It also has an active high clear (CLR) and a positive-edge-triggered clock (CLK) input. Switch IE should be up (1) and switches OE, CLK, and CLR should be down (0).

21. Click the On-Off switch to run the analysis. If the register output is not cleared, press the C key on the computer keyboard to raise the CLR switch to clear the register. Press the C key again to lower the CLR switch back to binary zero (0).

22. Set the data input switches (0–3) to any 4-bit binary number other than zero. Press the G key on the computer keyboard to lower the IE switch (active low input enable) to zero (0) to enable the input (D).

Question: Did the register output (Q3–Q0) change? Explain.

23. Press the space bar on the computer keyboard to make the CLK input rise from zero (0) to one (1) to produce a positive clock edge.

Questions: What did you observe at the register output (Q3-Q0)?

Is this register positive-edge-triggered or negative-edge-triggered?

Did the input data load in parallel or serial?

Is the data being read at the output in parallel or serial?

How many clock pulses were required to store the 4-bit binary input data?

24. Press the M key on the computer keyboard to raise the OE switch (active low output enable) to binary one (1).

Question: What did you observe at the register output (Q3–Q0)? Explain.

25. Press the M key again to lower switch OE (active low output enable) to zero (0). Press the G key to raise switch IE (active low input enable) to one (1) to disable the input (D). Change the data input switches (0–3) to a new 4-bit binary number. Press the space bar on the computer keyboard enough times to make it rise from zero (0) to one (1) and produce a positive clock edge.

Question: What did you observe at the register output (Q3–Q0)? Explain.

26. Click the On-Off switch to stop the analysis run. Pull down the File menu and open FIG24-6. You will test a 74194 universal 4-bit bidirectional shift register. This register has a parallel load data input (D–A) and two serial load data inputs (SL and SR). Serial load data input SL is a shift left serial input, and serial load data input SR is a shift right serial input. The mode control inputs (S1 S0) determine whether the register shifts left, shifts right, parallel loads, or stays the same when a positive clock edge is applied to the CLK input (See Table 24-1 in the Preparation section). Switch S provides the serial data inputs (SL and SR), switches D–A provide the parallel data inputs (D3–D0), and switches S1 and S0 control the mode inputs (S1 S0). The CLK and S switches should be down (0) and the CLR, S1, and S0 switches should be up (1).

27. Click the On-Off switch to run the analysis. If the register output is not cleared, press the R key on the computer keyboard to clear the register. Press the R key again to raise the CLR switch back to binary one (1).

28. With switches S1 and S0 up (S1 S0 = 11), set the data input switches (D–A) to any 4-bit binary number other than zero.

Question: Did the register output (Q3–Q0) change? Explain.

29. Press the space bar on the computer keyboard to make the CLK input rise from zero (0) to one (1) to produce a positive clock edge.

Questions: What did you observe at the register output (Q3–Q0)? Explain.

Is this register positive-edge-triggered or negative-edge-triggered?

Did the input data load in parallel or serial?

Is the data being read at the output in parallel or serial?

How many clock pulses were required to store the 4-bit binary input data?

30. Clear the register with the R key following the procedure in Step 27. Press the "1" key on the keyboard to set the mode control (S1 S0) to 01. Press the S key on the keyboard to raise the serial data input (S) to binary one (1). Press the space bar on the computer keyboard to make the CLK input fall from one (1) to zero (0) to produce a negative clock edge.

Questions: Did the register output (Q3–Q0) change?

Is this register positive-edge-triggered or negative-edge-triggered?

31. Press the space bar on the computer keyboard to make the CLK input rise from zero (0) to one (1) to produce a positive clock edge.

Questions: What did you observe at the register output (Q3–Q0)? Explain.

32. Press the space bar on the computer keyboard enough times to produce three more positive clock edges and observe the register output (Q3-Q0).

Questions: What did you observe at the register output (Q3–Q0)? Explain.

Did the input data load in parallel or serial? If serial, is it shift left or shift right?

Is the data being read at the output in parallel or serial?

How many clock pulses were required to store the 4-bit binary input data?

33. Clear the register with the R key following the procedure in Step 27. Press the "1" key and the "0" key on the keyboard to set the mode control (S1 S0) to 10. Press the space bar on the computer keyboard enough times to produce four positive clock edges and observe the register output (Q3–Q0).

Questions: What did you observe at the register output (Q3–Q0)? Explain.

Did the input data load in parallel or serial? If serial, is it shift left or shift right?

34. Press the "1" key on the keyboard to set the mode control (S1 S0) to 00. Press the S key on the keyboard to set the serial input (S) to zero (0). Press the space bar on the computer keyboard enough times to produce one or more positive clock edges and observe the register output (Q3–Q0).

Question: What did you observe at the register output (Q3–Q0)? Explain.

35. Click the On-Off switch to stop the analysis run. Pull down the File menu and open FIG24-7. You are looking at a circuit that will display the timing of a 74194 shift register. Switch S1 should be down (0) and switch S0 should be up (1). The word generator and logic analyzer settings should be as shown in Figure 24-7. Click the On-Off switch to run the analysis. The black curve plot on the logic analyzer screen is the clock (CLK) input to the register. The red curve plot is the serial data input (SL and SR). The blue (QA), dark green (QB), brown (QC), and light green (QD) curve plots represent the 4-bit register output (QD–QA).

NOTE: If this experiment is being performed in a hardwired laboratory, use two pulse generators in place of the word generator. If a logic analyzer is not available, use a dual-trace oscilloscope.

Question: Did the input data load in parallel or serial? If serial, is it shift left or shift right?

36. Click the On-Off switch to stop the analysis run. Press the "1" and "0" keys on the keyboard to set the mode switches (S1 S0) to 10. Click the On-Off switch to run the analysis again.

Question: Did the input data load in parallel or serial? If serial, is it shift left or shift right?

Name ______________________
Date ______________________

Asynchronous Counters

Objectives:

1. Demonstrate the application of J-K flip-flops to asynchronous (ripple) counters.
2. Demonstrate the effect of flip-flop propagation delay on asynchronous counter timing.
3. Investigate asynchronous (ripple) down counting.
4. Investigate the operation of a 7493 (74293) asynchronous (ripple) counter.
5. Demonstrate frequency division in asynchronous (ripple) counters.
6. Demonstrate how to change the modulus (divide-by) of an asynchronous (ripple) counter.

Materials:

One 5 V dc power supply
Four logic switches
Four logic probe lights
Four negative-edge-triggered J-K flip-flops (2-74112 ICs)
Two asynchronous (ripple) counters (2-7493 or 74293 ICs)
One function (pulse) generator
One logic analyzer or dual-trace oscilloscope

Preparation:

J-K flip-flops in the toggle mode (J=1, K=1) can be connected together to build a binary counter. The number of flip-flops determines how high the counter will count and the number of binary states (counts). The modulus (MOD) of a counter is equal to the number of binary states.

Counters are classified into two basic categories based on the way they are clocked. In asynchronous counters, the first flip-flop receives the input clock pulse and each successive flip-flop is clocked by the output of the preceding flip-flop. This is the reason that asynchronous counters are often referred to as ripple counters. In synchronous counters, all flip-flops receive the input clock pulse simultaneously. Synchronous counters will be studied in the next experiment.

Flip-flop propagation delay time (t_p) is the time it takes for a flip-flop output to change state after the flip-flop clock input receives a clock edge. Because each flip-flop in an asynchronous (ripple) counter is not clocked until the preceding flip-flop output changes state, all of the outputs do not respond to the counter clock input at the same time. Therefore, the time delay between the counter clock input and the response of the last flip-flop output depends on the number (N) of counter flip-flops. If the clock frequency is too high, the last flip-flop output will not change state before the next counter input clock pulse is applied. This will cause errors in the count. For this reason, the maximum clock frequency (f_{max}) for asynchronous (ripple) counters is lower than the maximum clock frequency for synchronous counters. For an asynchronous counter, the maximum clock frequency (f_{max}) can be calculated from

$$f_{max} = \frac{1}{(N \times t_p)}$$

Asynchronous (ripple) counters are often used as frequency dividers. Because J-K flip-flops that are in the toggle mode (J=1, K=1) change state only on each negative (or positive) edge of the clock pulse, the frequency of a flip-flop output waveform is one-half the frequency of the waveform at its clock input. Therefore, each flip-flop in an asynchronous (ripple) counter will divide the frequency output of the preceding flip-flop by two. Because the number of counter binary states (modulus) is multiplied by two for each additional flip-flop, the divide-by of the last counter flip-flop output is always equal to the modulus (MOD) of the counter.

The circuit in Figure 25-1 consists of four negative-edge-triggered J-K flip-flops wired as a 4-bit asynchronous (ripple) binary counter. Notice that the flip-flop J and K inputs are connected to 5 V (J=1, K=1), causing the flip-flops to toggle when they are clocked with a negative clock edge. The first flip-flop (Q0) CLK input is used as the counter CLK' input. Each successive flip-flop CLK input is connected to the preceding flip-flop output (Q). When the preceding flip-flop output (Q) drops from binary one (1) to binary zero (0), it will toggle the next flip-flop. The counter active low CLR' input is formed by connecting all of the flip-flop active low dc clear inputs together. When the CLR' input is dropped to binary zero (0), all of the flip-flops will clear (0), causing the counter to be cleared. The counter CLR' input must be returned to binary one (1) in order for the counter to count. If the active low CLR' input is not returned to binary one (1), the counter will stay cleared because the flip-flop dc clear inputs override the flip-flop CLK inputs.

The circuit in Figure 25-2 will demonstrate the effect of flip-flop propagation delay time (t_p) on the timing of a 4-bit asynchronous (ripple) counter. The word generator clock output will apply the clock pulses to the counter CLK' input. The word generator will also clear the counter by applying a zero (0) to the CLR' input on the first clock pulse, and then it will raise the CLR' input to one (1) for the rest of the count. The logic analyzer will monitor the counter clock input (CLK') and the counter outputs (Q3–Q0).

The circuit in Figure 25-3 consists of four negative-edge-triggered J-K flip-flops wired as a 4-bit asynchronous (ripple) binary down counter. Notice that each successive flip-flop CLK input is connected to the preceding flip-flop inverted output (Q'). This causes the counter to count down instead of counting up. The remainder of the circuit is the same as the up counter in Figure 25-1.

The 7493 (74293) in Figure 25-4 is an asynchronous (ripple) binary counter. It has four J-K flip-flops wired in the toggle mode, with flip-flop outputs QA, QB, QC, and QD. It is wired internally as a 1-bit (MOD-2 or divide-by-2) counter with negative-edge-triggered clock input CLKA and output QA, and a 3-bit (MOD-8 or divide-by-8) counter with negative-edge-triggered clock input CLKB and outputs QB, QC, and QD. The 7493 (74293) can be externally wired as a 4-bit (MOD-16 or divide-by-16) counter by connecting the output of the 1-bit counter (QA) to the clock input of the 3-bit counter (CLKB). This feature makes the 7493 (74293) very versatile because it can be wired as a 1-bit, 3-bit, or 4-bit counter. The 7493 (74293) also has two active high reset (clear) inputs (RO1 and RO2). These inputs can be used to reset (clear) the counter or to cut off the binary count in order to change the modulus (divide-by) of the counter. By wiring the counter in different configurations, the modulus (divide-by) of the counter can be changed to any value between MOD-2 (divide-by-2) and MOD-16 (divide-by-16).

The circuit in Figure 25-5 will display the timing of the 7493 (74293) asynchronous (ripple) counter. The word generator clock output will apply the clock pulses to the counter clock inputs. The logic analyzer will monitor the clock pulses and the counter outputs (QA–QD). When the D switch is down, the 7493 (74293) is configured as a 1-bit (MOD-2 or divide-by-2) counter using the CLKA clock input and a 3-bit (MOD-8 or divide-by-8) counter using the CLKB clock input. When switch D is up, the 7493 (74293) is configured as a 4-bit (MOD-16 or divide-by-16) counter using CLKA as the clock input with output QA connected to CLKB. The active high reset (clear) inputs (RO1 and RO2) are permanently connected to ground (0) to allow the counter to count.

The 7493 (74293) in Figure 25-6 is wired as a MOD-12 (divide-by-12) counter. Notice that it is wired as a 4-bit counter (QA connected to CLKB) and outputs QD and QC are connected to reset (clear) inputs RO1 and RO2. This will cause the counter to count between zero and eleven (12 states) and reset (clear) on the twelfth clock pulse (binary output 1100), making it a MOD-12 (divide-by-12) counter.

The 7493 (74293) counters in Figures 25-7, 25-8, and 25-9 are connected as divide-by counters. The clock input and the counter output will be displayed on the oscilloscope screen. You can determine the expected modulus (divide-by) of these counters by observing the wiring of QA, CLKB, RO1, and RO2 and following the reasoning used in the discussion for Figure 25-6. You can measure the modulus (divide-by) of these counters by measuring the time period for one cycle of the clock pulse (T_c) and the time period for one cycle of the counter output (T_o). From these values, you can calculate the clock frequency (f_c) and the counter output frequency (f_o) using the equation

$$f = \frac{1}{T}$$

The modulus (divide-by) is calculated from

$$\text{Modulus} = \frac{f_c}{f_o}$$

The circuit in Figure 25-10 consists of two 7493 (74293) cascaded asynchronous counters to increase the modulus (divide-by). The function generator will apply a 100 kHz pulse frequency to the clock input. The oscilloscope will monitor the cascaded counter output. You can determine the expected modulus (divide-by) of the cascaded counters by first determining the expected modulus (divide-by) of each of the two counters. The MOD number (divide-by) of the cascaded pair is equal to the product of their individual MOD numbers. The output frequency (f_o) is then determined by dividing the input clock frequency by the modulus (divide-by) of the cascaded pair.

Figure 25-1 4-Bit Asynchronous (Ripple) Counter

Figure 25-2a 4-Bit Asynchronous (Ripple) Counter Timing (EWB Version 4)

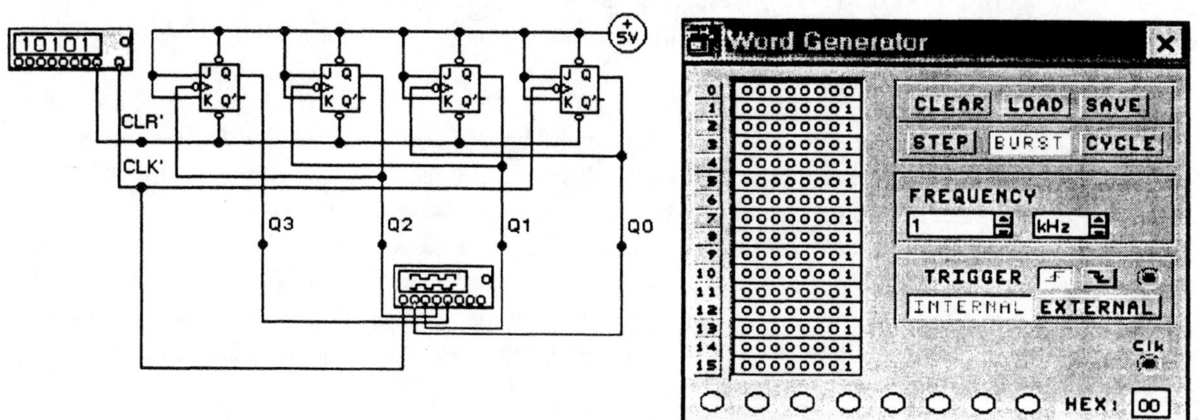

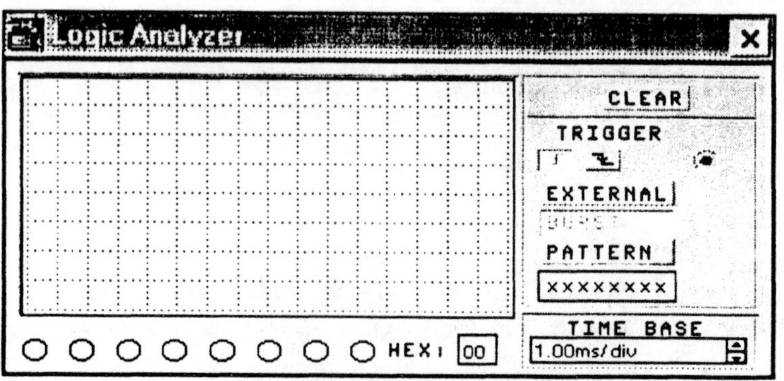

Figure 25-2b 4-Bit Asynchronous (Ripple) Counter Timing (EWB Version 5)

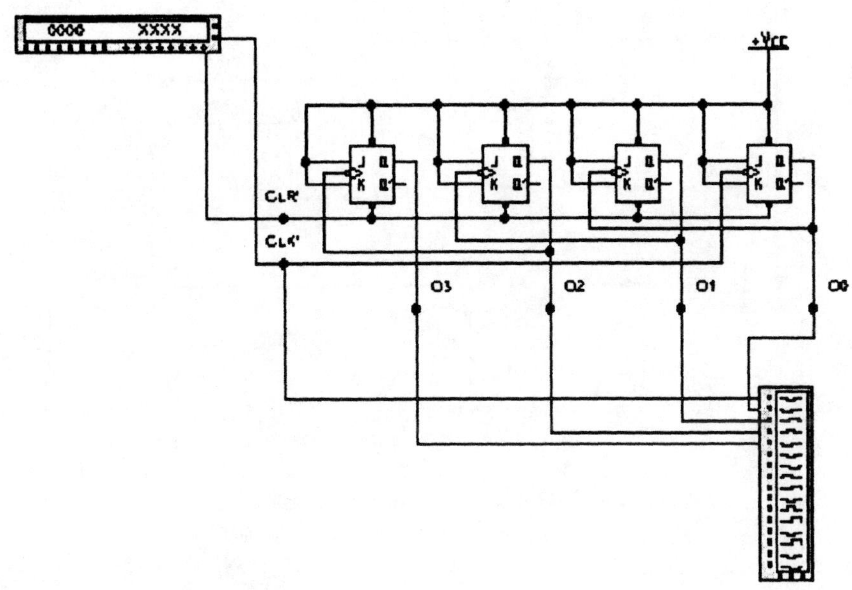

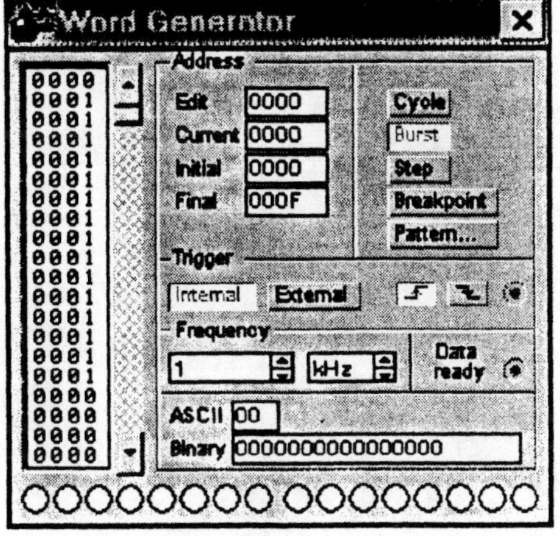

Logic Analyzer Settings
Clocks per division ---------------8

Clock Setup dialog box
Clock edge -------------- positive
Clock mode ------------- Internal
Internal clock rate ------ 10 kHz
Clock qualifier ---------- x
Pre-trigger samples --- 100
Post-trigger samples -- 1000
Threshold voltage (V) 2

Trigger Patterns dialog box
A ---------------------------- xxxxxxxxxxxxxxxx
Trigger combinations -- A
Trigger qualifier --------- x

Figure 25-3 4-Bit Asynchronous (Ripple) Down Counter

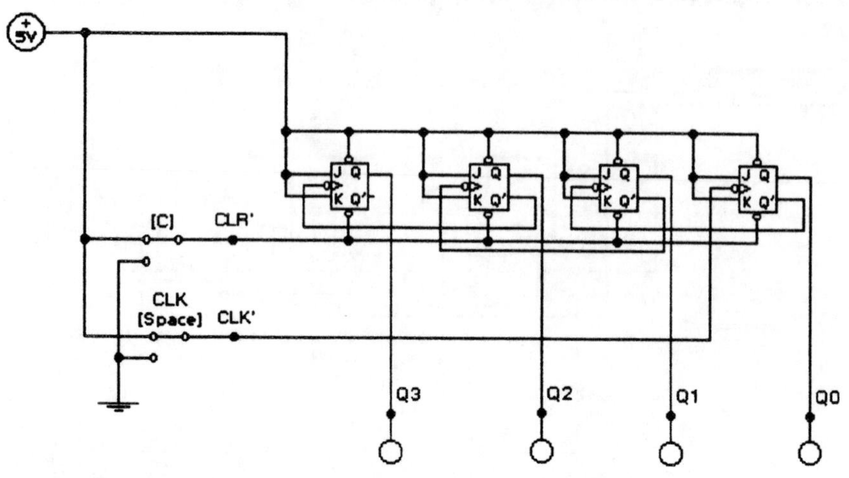

Figure 25-4 7493 Asynchronous (Ripple) Counter

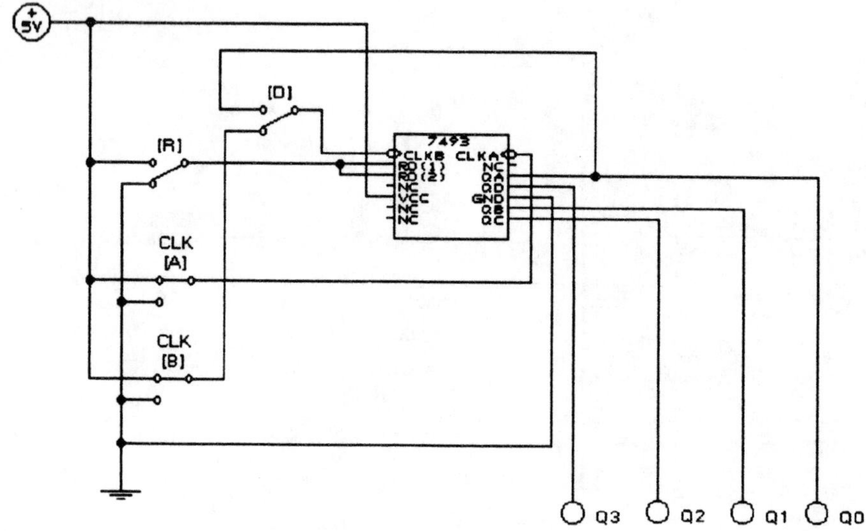

Figure 25-5a 7493 Asynchronous (Ripple) Counter Timing (EWB Version 4)

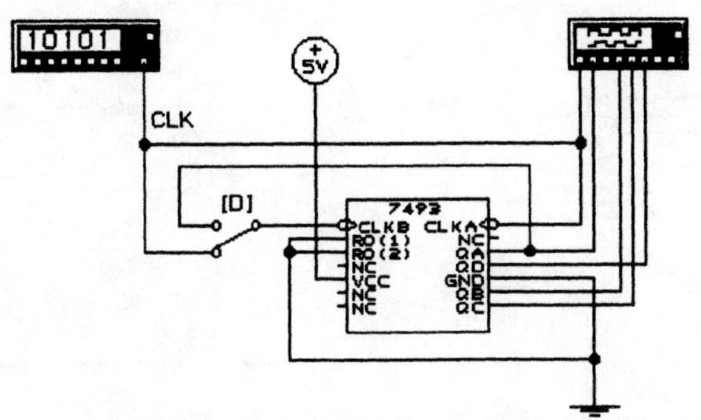

Figure 25-5b 7493 Asynchronous (Ripple) Counter Timing (EWB Version 5)

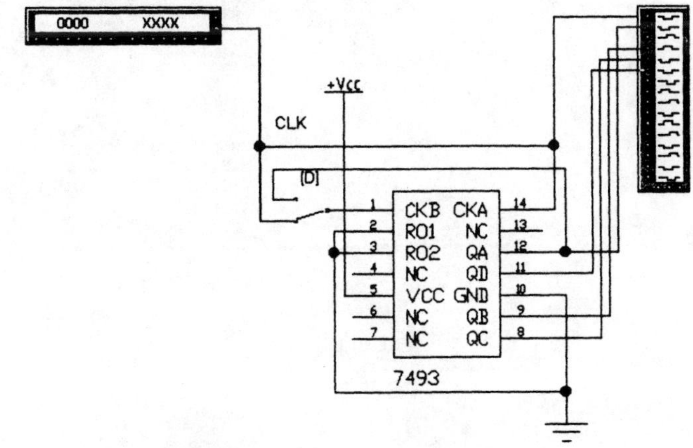

Figure 25-6 7493 Wired as a MOD-12 Counter

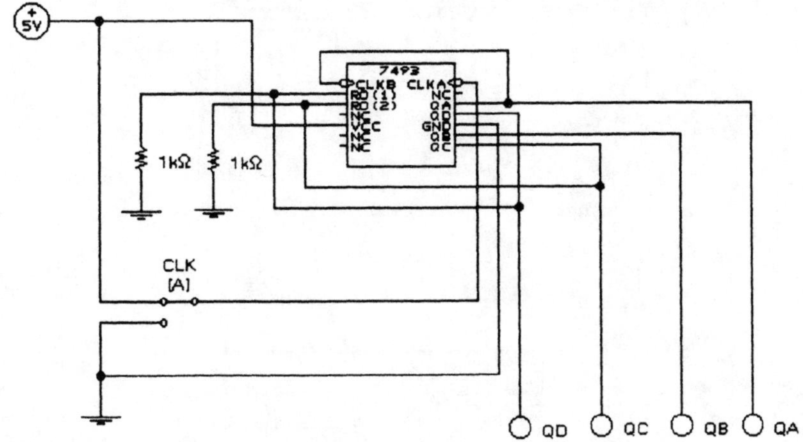

Figure 25-7 7493 Frequency Division

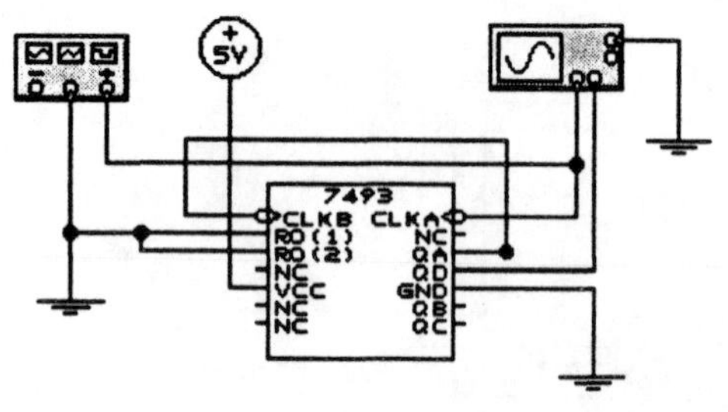

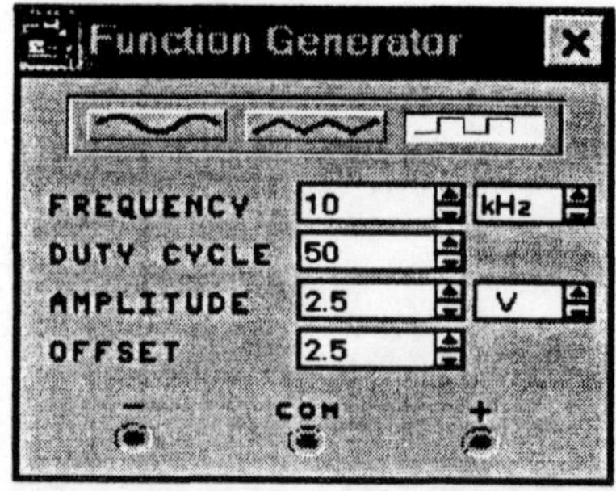

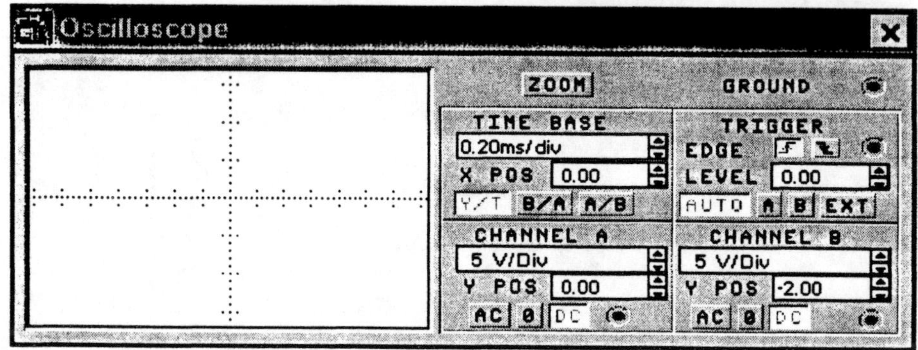

Figure 25-8 7493 Frequency Division

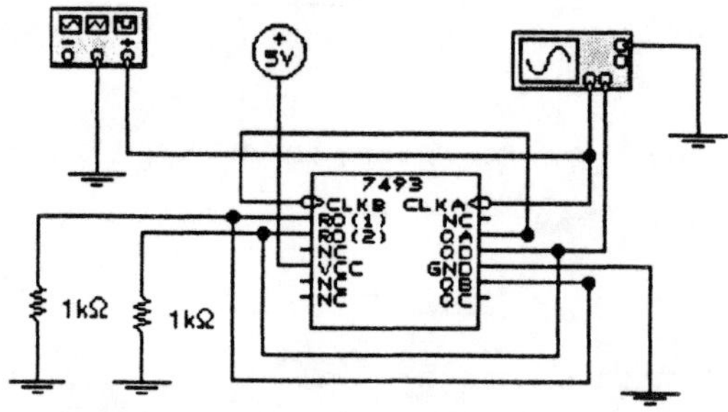

Figure 25-9 7493 Frequency Division

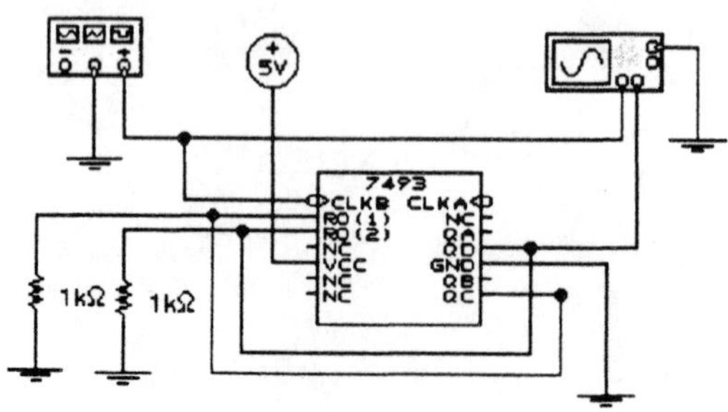

Figure 25-10 7493 Frequency Division

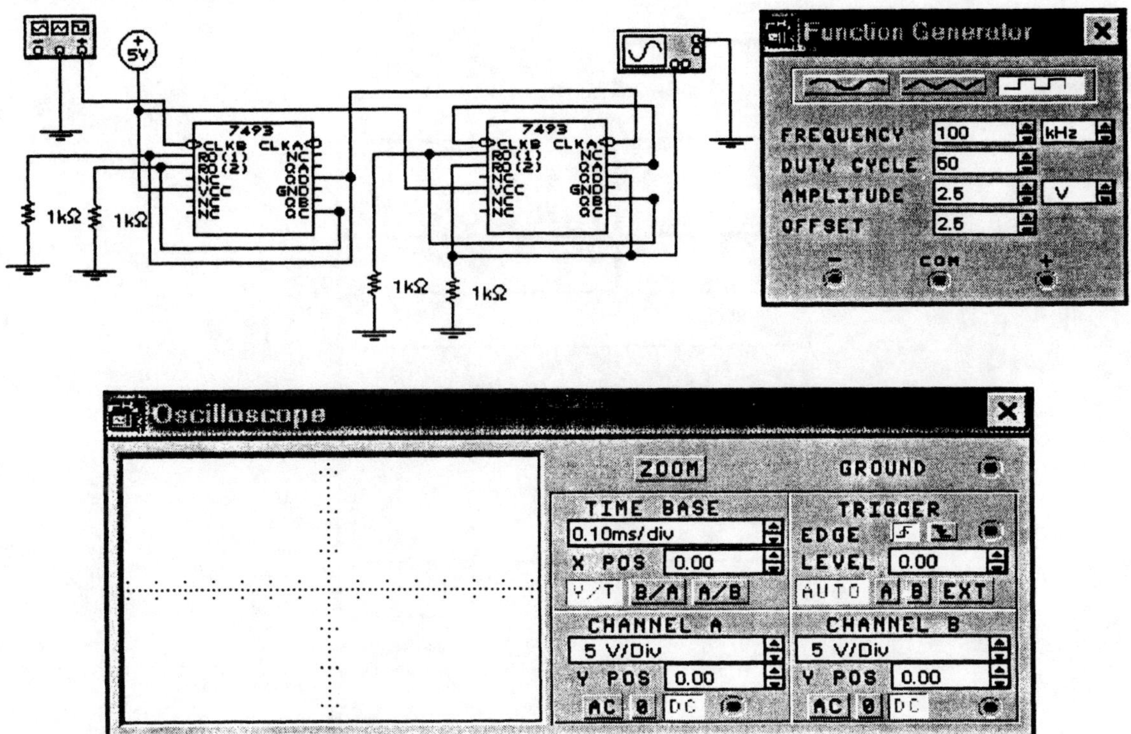

Procedure:

1. Pull down the File menu and open FIG25-1. You are looking at four negative-edge-triggered J-K flip-flops wired as a 4-bit asynchronous (ripple) binary counter. The C and CLK switches should be up (1). Click the On-Off switch to run the analysis and press the C key on the computer keyboard to clear the counter. Press the C key again to raise CLR' back to binary one (1).

NOTE: If this experiment is being performed in a hardwired laboratory, use two 74112 ICs for the four negative-edge-triggered J-K flip-flops.

Questions: Is the clear input (CLR') active low or active high?

Why does the CLR' input need to be binary one (1) in order for the counter to be able to count?

2. Press the space bar on the computer keyboard to make the CLK' input drop from one (1) to zero (0) to produce a negative clock edge. Record the counter binary output for one clock pulse in Table 25-1.

Table 25-1 Output Count

Clock Pulse	Output Q3 Q2 Q1 Q0
1	
2	
3	
4	
5	
6	
7	
8	
9	
10	
11	
12	
13	
14	
15	
16	

Question: Is the counter positive-edge-triggered or negative-edge-triggered?

3. Press the space bar on the computer keyboard enough times to produce another negative clock edge. Record the counter binary output for two clock pulses in Table 25-1. Repeat this procedure until the table is complete.

Questions: Based on the data in Table 25-1, what conclusion can you draw about the relationship between the counter binary output and the number of clock pulses applied to the counter CLK' input?

What happened when the counter was clocked after the maximum count was reached?

4. Click the On-Off switch to stop the analysis run. Pull down the File menu and open FIG25-2. You are looking at a circuit that will demonstrate the effect of flip-flop propagation delay time (t_p) on the timing of a 4-bit asynchronous (ripple) counter. The word generator and logic analyzer settings should be as shown in Figure 25-2. Click the On-Off switch to run the analysis. The black curve plot on the logic analyzer screen is the clock (CLK') input to the counter. The blue (Q0), dark green (Q1), brown (Q2), and light green (Q3) curve plots represent the 4-bit counter output (Q3–Q0). Draw and label the CLK' input and the counter output waveshapes in the space provided.

NOTE: If this experiment is being performed in a hardwired laboratory, use two 74112 ICs for the four negative-edge-triggered J-K flip-flops and use a pulse generator in place of the word generator. If a logic analyzer is not available, use a dual-trace oscilloscope.

Questions: Is each output changing state on the positive or the negative edge of the previous flip-flop output pulse? Explain.

What is the relationship between the frequencies of the flip-flop output pulses?

What is the relationship between the frequency of each flip-flop output pulse and the CLK' input pulse?

Are the output binary counts correct after each clock input pulse (negative edge)?

5. Click the On-Off switch to stop the analysis run. Change the frequency of the word generator to 50 MHz. Change the time base on the logic analyzer to 0.02 μsec/div (Version 4). Change the internal clock rate on the logic analyzer to 500 MHz (Version 5). Click the On-Off switch to run the analysis. Draw and label the new CLK' input and counter output waveshapes in the space provided.

Questions: Are all of the counter output pulses changing state on the negative edge of the previous flip-flop output pulse? Explain.

Are the output binary counts still correct after all of the clock inputs (negative edges)? Explain.

How does the propagation delay time (t_p) for each flip-flop compare with the time period (T) of the clock pulses?

Based on the curve plots, what is the approximate propagation delay time (t_p) for the J-K flip-flops in this asynchronous counter? (The clock time period (T), the inverse of the clock frequency, is equal to .02 μsec).

6. Calculate the counter maximum clock frequency (f_{max}) based on the flip-flop propagation delay time (t_p) and the number of counter flip-flops (N)?

7. Click the On-Off switch to stop the analysis run. Pull down the File menu and open FIG25-3. You are looking at four negative-edge-triggered J-K flip-flops wired as a 4-bit asynchronous (ripple) binary down counter. The C and CLK switches should be up (1). Click the On-Off switch to run the analysis and press the C key on the computer keyboard to clear the counter. Press the C key again to raise CLR' back to binary one (1).

> NOTE: If this experiment is being performed in a hardwired laboratory, use two 74112 ICs for the four negative-edge-triggered J-K flip-flops.

8. Press the space bar on the computer keyboard to make the CLK' input drop from one (1) to zero (0) to produce a negative clock edge. Record the counter binary output for one clock pulse in Table 25-2. Press the space bar on the computer keyboard enough times to produce another negative clock edge. Record the counter binary output for two clock pulses in Table 25-2. Repeat this procedure until the table is complete.

Table 25-2 Output Count

Clock Pulse	Output Q3 Q2 Q1 Q0
1	
2	
3	
4	
5	
6	
7	
8	
9	
10	
11	
12	
13	
14	
15	
16	

Question: Based on the data in Table 25-2, what conclusion can you draw about the direction of the binary count as the clock pulses are applied to the counter CLK' input?

9. Click the On-Off switch to stop the analysis run. Pull down the File menu and open FIG25-4. You will test a 7493 (74293) asynchronous (ripple) counter. This counter has two active high reset (clear) inputs (RO1 and RO2) and two negative-edge-triggered clock inputs (CLKA and CLKB). The CLKA input is used to trigger flip-flop A and the CLKB input is used to trigger the 3-bit counter using flip-flops B, C, and D. The 7493 (74293) can be externally wired as a 4-bit (MOD-16) counter by connecting flip-flop output QA to clock input CLKB and using CLKA as the clock input. When the D switch is down, the 7493 (74293) is configured as a 1-bit (MOD-2) counter using the CLKA input and a 3-bit (MOD-8) counter using the CLKB input. When the D switch is up, the 7493 (74293) is configured

as a 4-bit (MOD-16) counter using the CLKA input. See the Preparation section for more details about the 7493 (74293) asynchronous (ripple) counter.

10. Switches R and D should be down (0) and the CLKA and CLKB switches should be up (1). Click the On-Off switch to run the analysis. If the counter output (Q3–Q0) is not cleared, press the R key on the computer keyboard to raise the reset (clear) input (RO1 and RO2) to binary one (1) to clear the counter. Press the R key again to lower the reset input back to binary zero (0).

11. With switch D down, press the A key on the computer keyboard to produce a negative clock edge on the CLKA input. Keep pressing the A key to produce a number of negative clock edges on the CLKA input.

Question: What did you observe at counter output Q0? Explain.

12. With switch D down, press the B key on the computer keyboard to produce a negative clock edge on the CLKB input. Keep pressing the B key to produce a number of negative clock edges on the CLKB input.

Question: What did you observe at the counter outputs (Q3–Q1)? Explain.

13. Press the D key on the computer keyboard to raise switch D up. This will connect output QA to input CLKB. Press the R key to reset (clear) the counter, and then press it again to lower it to zero (0). Keep pressing the A key to apply a number of negative clock edges to the CLKA input and observe the output (Q3–Q0).

Question: What did you observe at the counter outputs (Q3–Q0)? Explain.

14. Click the On-Off switch to stop the analysis run. Pull down the File menu and open FIG25-5. You are looking at a circuit that will display the timing of a 7493 (74293) asynchronous (ripple) counter. The word generator and logic analyzer settings should be as shown in Figure 25-2. When the D switch is down, the 7493 (74293) is configured as a 1-bit (MOD-2) counter using the CLKA input and a 3-bit (MOD-8) counter using the CLKB input. When the D switch is up, the 7493 (74293) is configured as a 4-bit (MOD-16) counter using the CLKA input. Switch D should be down.

> NOTE: If this experiment is being performed in a hardwired laboratory, use a pulse generator in place of the word generator. If a logic analyzer is not available, use a dual-trace oscilloscope.

15. Click the On-Off switch to run the analysis. The black curve plot on the logic analyzer screen is the clock input (CLKA and CLKB). The blue curve plot represents the 1-bit (MOD-2) counter output (QA) clocked on the CLKA input. The dark green (QB), brown (QC), and light green (QD) curve plots represent the 3-bit (MOD-8) counter output clocked on the CLKB input. Draw and label the curve plots in the space provided.

Questions: What is the frequency relationship between the CLKA input (black) and the counter QA output (blue)? What is the modulus (divide-by) of this counter?

What is the frequency relationship between the CLKB input (black) and the counter outputs QB (dark green), QC (brown), and QD (light green)? What is the modulus (divide-by) of this counter?

16. Click the On-Off switch to stop the analysis run. Press the D key on the computer keyboard to raise switch D. This will connect output QA to input CLKB. Click the On-Off switch to run the analysis again. Draw and label the new curve plots in the space provided.

Question: What is the frequency relationship between the CLKA input (black) and counter outputs QA (blue), QB (dark green), QC (brown), and QD (light green)? What is the modulus (divide-by) of this counter?

17. Click the On-Off switch to stop the analysis run. Pull down the File menu and open FIG25-6. You will test a 7493 (74293) wired as a MOD-12 (divide-by-12) counter. Notice that output QA is connected to the CLKB input, output QD is connected to reset input RO1, and output QC is connected to reset input RO2. Click the On-Off switch to run the analysis. Press the A key on the computer keyboard enough times to apply a series of negative clock edges to the CLKA input and observe the output (QD–QA). Record the results in Table 25-3.

Table 25-3 Output Count

Clock Pulse	Output QD QC QB QA
1	
2	
3	
4	
5	
6	
7	
8	
9	
10	
11	
12	
13	
14	
15	
16	

Questions: What conclusion can you draw about the relationship between the counter binary output and the number of clock pulses applied to the CLKA input?

What happened on clock pulse number twelve? Explain.

Based on these results, what is the modulus (divide-by) of this counter? Does it match the expected value based on the circuit configuration?

18. Click the On-Off switch to stop the analysis run. Pull down the File menu and open FIG25-7. The function generator and oscilloscope settings should be as shown in Figure 25-7. Click the On-Off switch to run the analysis. Use the "zoom" feature on the oscilloscope to determine the time period (T) for one cycle and the frequency (f) of the clock input (black) and the counter output (blue). Record your answers in the space provided.

Clock input T_c = ____________

f_c = ____________

Counter output T_o = ____________

f_o = ____________

19. Based on the results in Step 18, calculate the modulus (divide-by) of this counter.

Questions: Based on the circuit configuration in Figure 25-7, what is the expected modulus (divide-by) of this counter? Explain.

How does this answer compare with the calculated modulus in Step 19?

20. Click the On-Off switch to stop the analysis run. Pull down the File menu and open FIG25-8. The function generator and oscilloscope settings should be as shown in Figure 25-7. Click the On-Off switch to run the analysis. Use the "zoom" feature on the oscilloscope to determine the time period (T) for one cycle and the frequency (f) of the clock input (black) and the counter output (blue). Record your answers in the space provided.

Clock input T_c = ____________

f_c = ____________

Counter output T_o = ____________

f_o = ____________

21. Based on the results in Step 20, calculate the modulus (divide-by) of this counter.

Questions: Based on the circuit configuration in Figure 25-8, what is the expected modulus (divide-by) of this counter? Explain.

How does this answer compare with the calculated modulus in Step 21?

22. Click the On-Off switch to stop the analysis run. Pull down the File menu and open FIG25-9. The function generator and oscilloscope settings should be as shown in Figure 25-7. Click the On-Off switch to run the analysis. Use the "zoom" feature on the oscilloscope to determine the time period (T) for one cycle and the frequency (f) of the clock input (black) and the counter output (blue). Record your answers in the space provided.

Clock input T_c = ____________

f_c = ____________

Counter output T_o = ____________

f_o = ____________

23. Based on the results in Step 22, calculate the modulus (divide-by) of this counter.

Questions: Based on the circuit configuration in Figure 25-9, what is the expected modulus (divide-by) of this counter? Explain.

How does this answer compare with the calculated modulus in Step 23?

24. Click the On-Off switch to stop the analysis run. Pull down the File menu and open FIG25-10. The function generator and oscilloscope settings should be as shown in Figure 25-10. Click the On-Off switch to run the analysis. Use the "zoom" feature on the oscilloscope to determine the time period (T_o) for one cycle and the frequency (f_o) of the counter output (blue). Record your answers in the space provided.

T_o = ____________

f_o = ____________

25. Based on the cascaded counter circuit in Figure 25-10 and the 100 kHz clock frequency (f_c), calculate the expected output frequency (f_o).

Question: How did your calculated output frequency (f_o) compare with the value measured in Step 24?

Name ______________________

Date ______________________

Synchronous Counters

Objectives:

1. Demonstrate the application of J-K flip-flops to synchronous binary counters.
2. Demonstrate the effect of propagation delay on synchronous counter timing.
3. Investigate synchronous up/down counting.
4. Investigate the operation of the 74191 presettable up/down synchronous binary counter.
5. Demonstrate frequency division in synchronous counters.
6. Demonstrate how to change the modulus (divide-by) of a synchronous counter.
7. Investigate cascaded synchronous counters.

Materials:

One 5 V dc power supply
Eight logic switches
Ten logic probe lights
Four negative-edge-triggered J-K flip-flops (2-74112 ICs)
Two presettable up/down synchronous binary counters (2-74191 ICs)
Four two-input AND gates (1-7408 IC)
Two two-input OR gates (1-7432 IC)
One INVERTER (1-7404 IC)
One function (pulse) generator
One logic analyzer or dual-trace oscilloscope

Preparation:

Make sure you complete Experiment 25 before attempting this experiment. Also, review the Preparation section of Experiment 25.

Synchronous counters eliminate the problem of accumulated propagation delay encountered in asynchronous counters because all of the flip-flops receive the input clock pulses simultaneously in synchronous counters. Therefore, some means must be used to control when a flip-flop will toggle and when it will not toggle in a synchronous counter. This is accomplished with logic networks connected between the J-K flip-flop inputs and the previous flip-flop outputs. Due to the additional logic circuitry, synchronous counters have more complicated circuitry and higher cost than asynchronous counters.

Because of the elimination of the accumulated propagation delay time (t_p) in synchronous counters, the time delay between the counter clock input and the response of the last flip-flop output does not depend on the number of counter flip-flops. For this reason, the maximum clock frequency (f_{max}) for synchronous counters is higher than the maximum clock frequency for asynchronous counters. For a synchronous counter, the maximum clock frequency can be calculated from

$$f_{max} = \frac{1}{(\text{FF } t_p + \text{AND gate } t_p)}$$

Synchronous counters are also often used as frequency dividers. Each output in a synchronous counter is equal to one-half the frequency of the preceding output. Because the number of synchronous counter binary states (modulus) is multiplied by two for each additional output, the divide-by of the last counter output is always equal to the modulus (MOD) of the counter.

The circuit in Figure 26-1 consists of four negative-edge-triggered J-K flip-flops wired as a 4-bit synchronous binary counter. Notice that all of the flip-flop clock inputs are connected to the counter CLK' input. Each flip-flop J and K input is connected to either 5 V (J=1, K=1), the output of another flip-flop, or the output of a logic gate (AND gate). The correct logic causes the flip-flops to toggle at the proper time to produce a binary up count. The counter active low CLR' input is formed by connecting all of the flip-flop active low dc clear inputs together. When the CLR' input is dropped to binary zero (0), all of the flip-flops will clear (0), causing the counter to be cleared. The counter CLR' input must be returned to binary one (1) in order for the counter to count. If the active low CLR' input is not returned to binary one (1), the counter will stay cleared because the flip-flop dc clear inputs override the flip-flop CLK inputs.

The circuit in Figure 26-2 will demonstrate the effect of flip-flop and AND gate propagation delay time (t_p) on the timing of a 4-bit synchronous binary counter. The word generator will apply the clock pulses to the counter CLK' input. The word generator will also clear the counter by applying a zero (0) to the CLR' input on the first clock pulse, and then it will raise the CLR' input to one (1) for the rest of the count. The logic analyzer will monitor the clock input (CLK') and the counter outputs (Q3–Q0).

The circuit in Figure 26-3 consists of three negative-edge-triggered J-K flip-flops wired as a 3-bit synchronous up/down binary counter. Notice that the down count logic circuitry (bottom AND gates) is identical to the up count logic circuitry (top AND gates), except the down count logic is connected to the inverted flip-flop outputs (Q'). When the UP/DOWN' switch is up, the up count logic gates are opened and the down count logic gates are closed. When the UP/DOWN' switch is down, the up count logic gates are closed and the down count logic gates are opened.

The 74191 in Figure 26-4 is a presettable up/down synchronous binary counter. It can be preset to any count by applying a 4-bit binary number to inputs DCBA and dropping the active low load input (LOAD) to binary zero (0). The load input (LOAD) must be returned to binary one (1) in order for the counter to count because the load input overrides the CLK input. The 74191 will count up if a binary zero (0) is applied to the U/D input and it will count down if a binary one (1) is applied to the U/D input. The counter will increment or decrement one count each time the CLK input receives a positive clock edge. The active low count enable input (CTEN) must be low (0) for the counter to be enabled, otherwise it will be disabled. The TC (MAX/MIN) output is normally low (0), but will go high (1) for one clock period when the counter reaches zero (0000) in the count down mode or fifteen (1111) in the count up mode. The RC (RCO) output is normally high (1), but will go low (0) for the low half of the clock period when the counter reaches zero (0000) in the count down mode or fifteen (1111) in the count up mode. The TC (MAX/MIN) and RC (RCO) outputs are made available in order to be able to cascade 74191 synchronous counters to provide a count higher than fifteen (1111). Each additional cascaded counter adds 4 bits to the output.

The circuit in Figure 26-5 will display the timing of the 74191 synchronous binary counter. The word generator will apply the clock pulses to the counter clock input. It will also apply a binary zero (0) to the load input (LOAD) on the first clock pulse and a binary one (1) for the rest of the count. This will cause the counter to start the count at the binary number set by the DCBA switches. The logic analyzer will monitor the clock pulses, the load input (LOAD), the counter outputs (QA–QD), the ripple clock output (RCO), and the terminal count output (MAX/MIN).

The 74191 in Figure 26-6 is wired as a MOD-12 (divide-by-12) counter. Notice that the MAX/MIN output and the CLK input are connected to a NAND gate that feeds the LOAD input. This will cause the counter to reset the count during the positive half of the clock period when the counter output reaches the terminal state and the MAX/MIN output is high (1). The counter will reset to the binary count that is set by the DCBA switches. (Up equals a binary one and down equals a binary zero). Because this counter is in the count down mode, the binary number represented by the DCBA switches will determine the modulus (number of binary states) of the counter. Therefore, the modulus (MOD) of this counter can be changed by simply changing the DCBA switches, eliminating the need to change the wiring.

The 74191 synchronous counter in Figure 26-7 is connected as a divide-by counter. The clock input and the counter output will be displayed on the oscilloscope screen. You can determine the expected modulus (divide-by) of this counter by observing the position of the DCBA switches (up equals binary one and down equals binary zero) and following the reasoning used in the discussion for Figure 26-6. You can measure the modulus (divide-by) of this counter by measuring the time period for one cycle of the clock pulse (T_c) and the time period for one cycle of the counter output (T_o). From these values, you can calculate the clock frequency (f_c) and the counter output frequency (f_o) using the equation

$$f = \frac{1}{T}$$

The modulus (divide-by) is calculated from

$$\text{Modulus} = \frac{f_c}{f_o}$$

The circuit in Figure 26-8 consists of two 74191 cascaded synchronous counters wired as an 8-bit counter. The counter on the right outputs the low 4-bits and the counter on the left outputs the high 4-bits. The cascaded 8-bit synchronous counter can count up or down depending on the position of the UP'/DOWN switch. The RCO output of the counter on the right is connected to the active low enable input (CTEN) of the counter on the left. This will cause the counter on the left to be enabled during the low half of the clock period and increment or decrement one count when the counter on the right reaches its terminal count and begins a new count cycle. Notice that both counters are being clocked at the same time, making the cascaded pair synchronous.

Figure 26-1 4-Bit Synchronous Counter

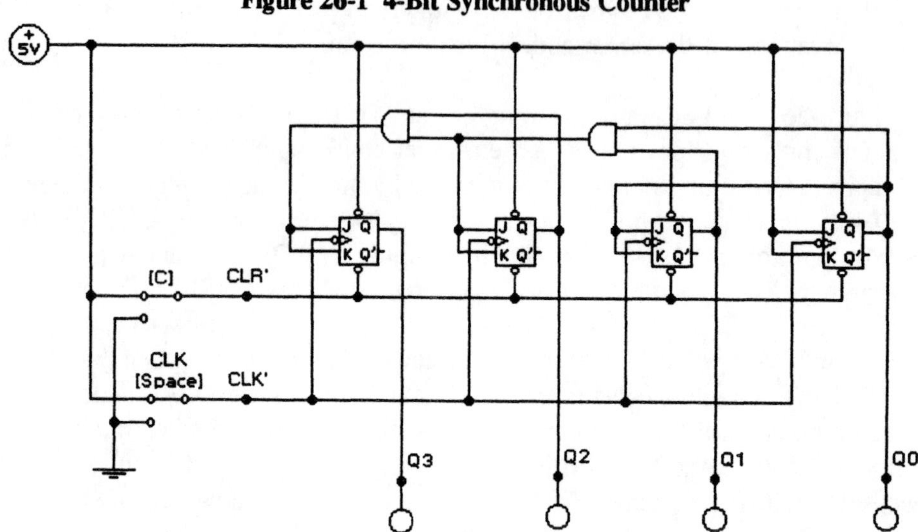

Figure 26-2a 4-Bit Synchronous Counter Timing (EWB Version 4)

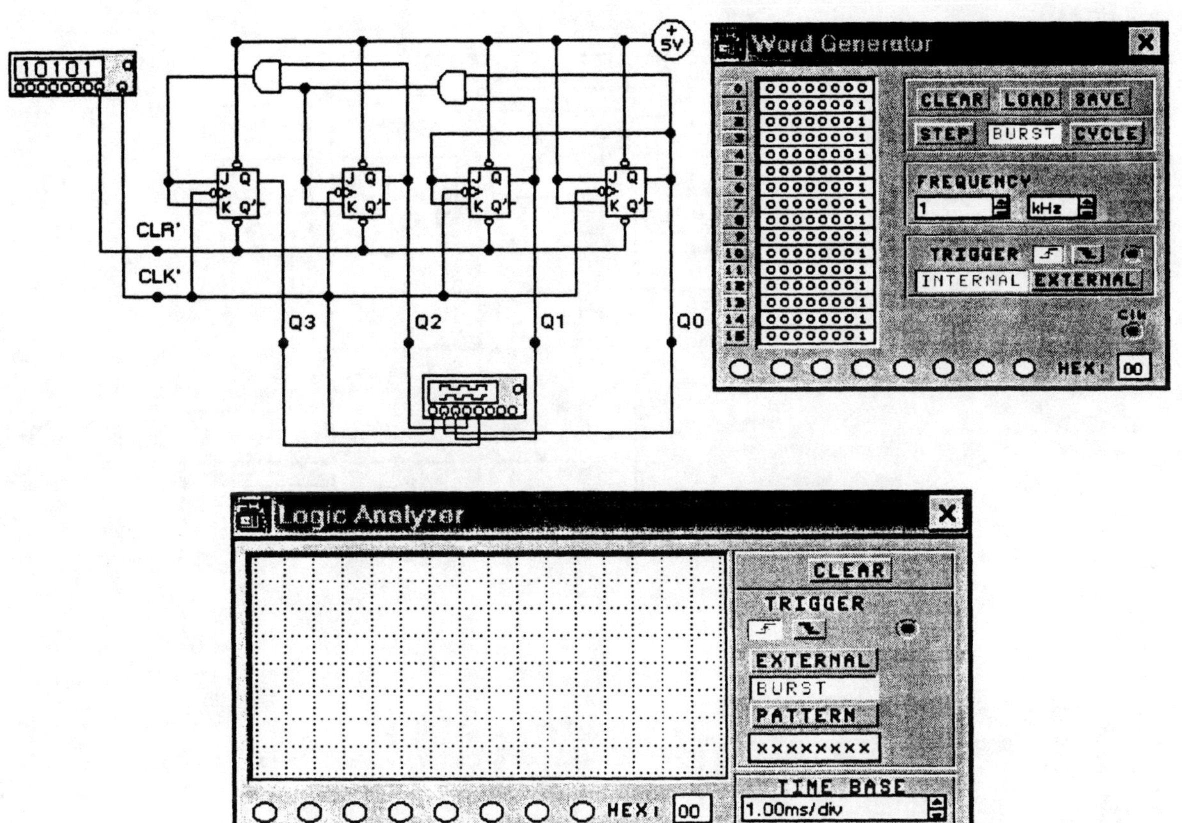

Figure 26-2b 4-Bit Synchronous Counter Timing (EWB Version 5)

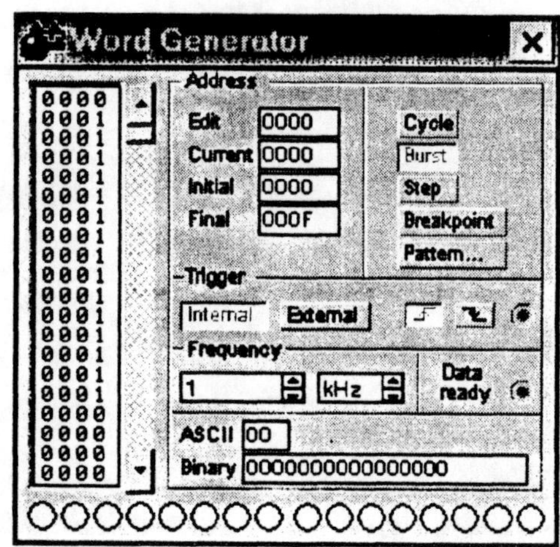

Logic Analyzer Settings
Clocks per division ---------------- 8

Clock Setup dialog box
Clock edge --------------- positive
Clock mode -------------- Internal
Internal clock rate ------ 10 kHz
Clock qualifier ----------- x
Pre-trigger samples --- 100
Post-trigger samples -- 1000
Threshold voltage (V) 2

Trigger Patterns dialog box
A ------------------------------ xxxxxxxxxxxxxxxx
Trigger combinations -- A
Trigger qualifier ---------- x

Figure 26-3 3-Bit Synchronous Up/Down Counter

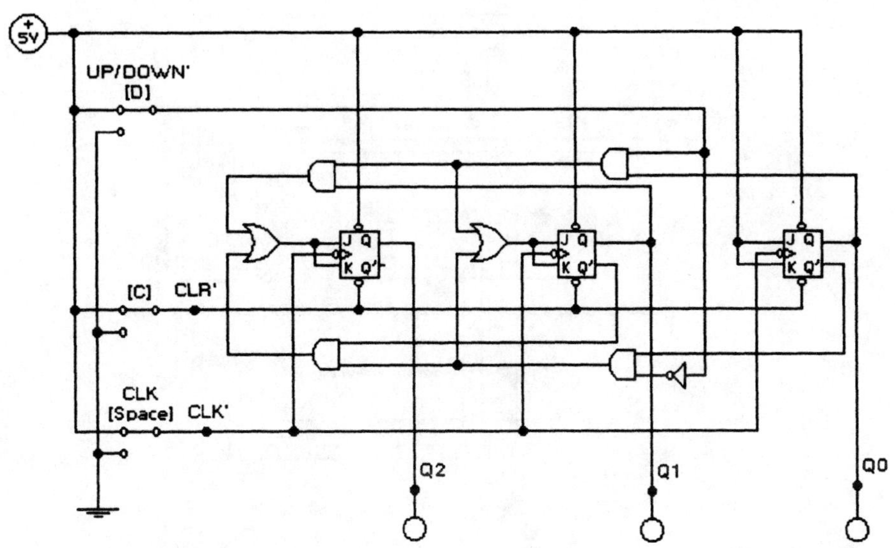

Figure 26-4 74191 Presettable Up/Down Synchronous Binary Counter

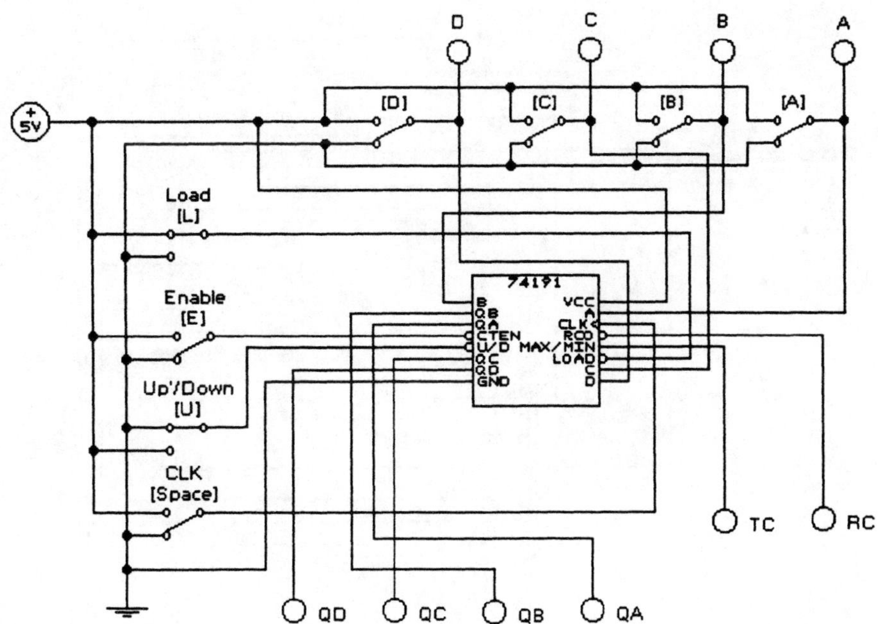

Figure 26-5a 74191 Synchronous Counter Timing (EWB Version 4)

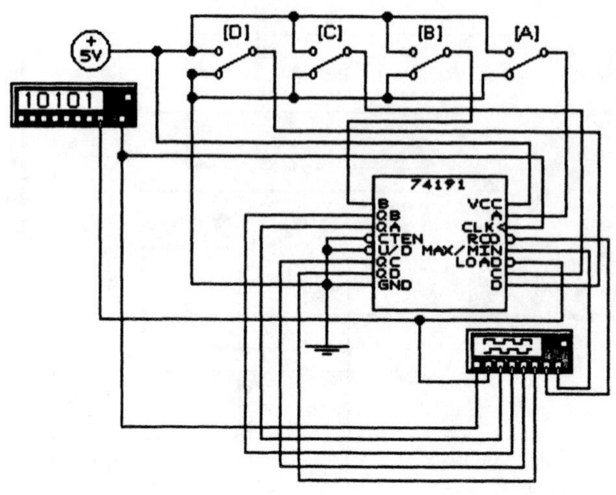

Figure 26-5b 74191 Synchronous Counter Timing (EWB Version 5)

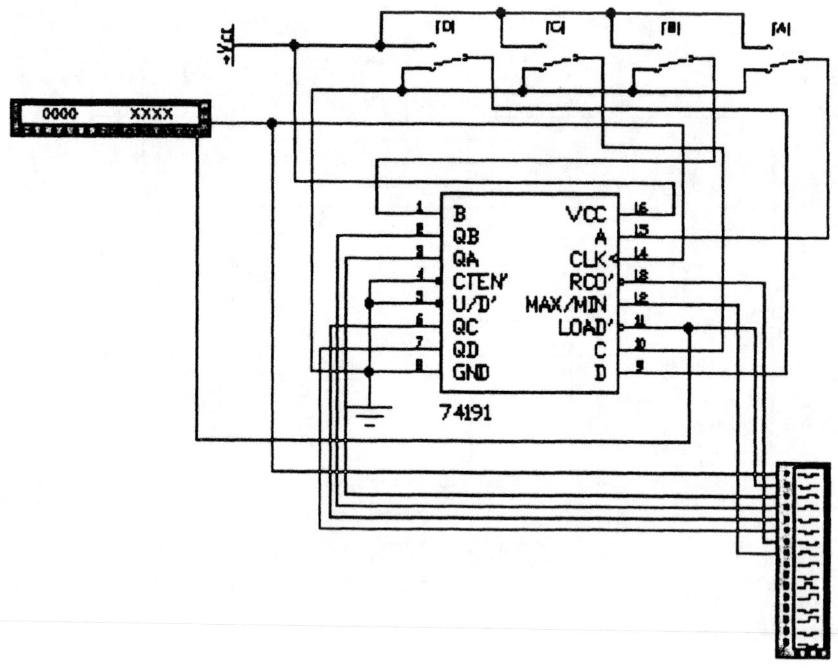

Figure 26-6 74191 Wired as a MOD-12 Counter

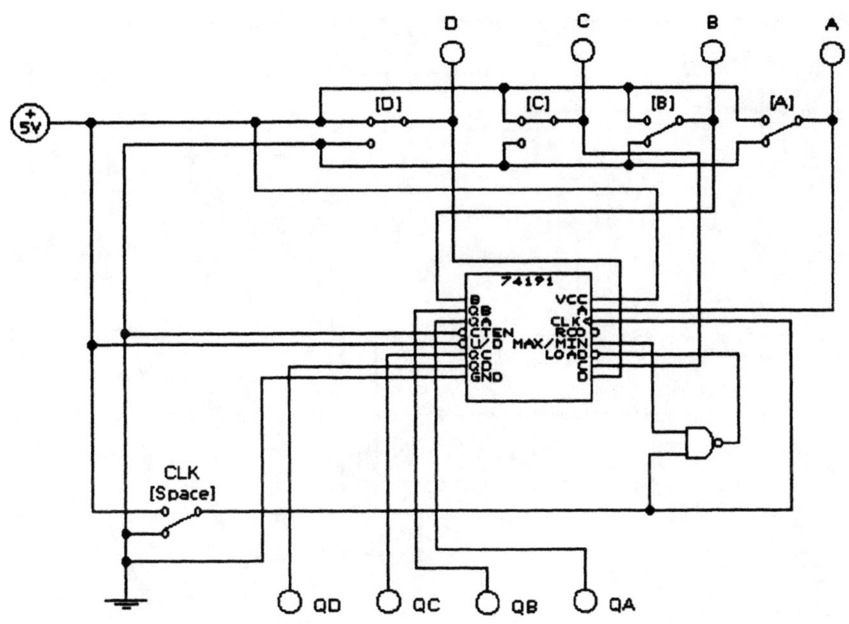

Figure 26-7 74191 Frequency Division

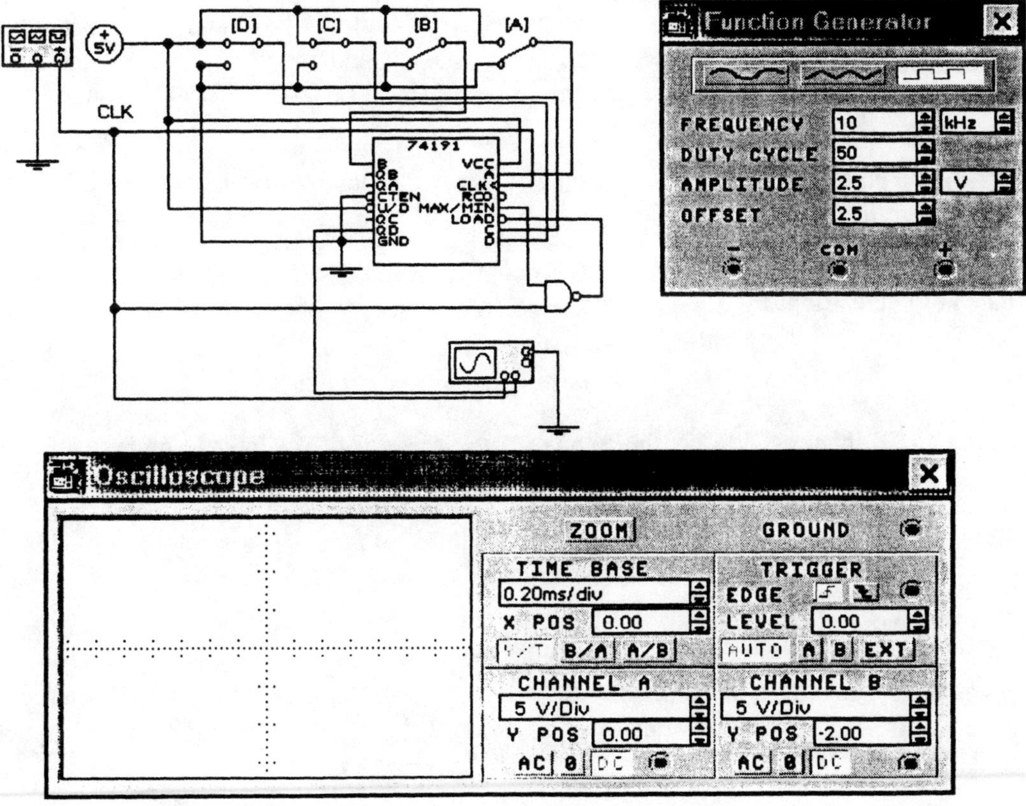

Figure 26-8 Cascaded 74191 Synchronous Counters (8-Bit Counter)

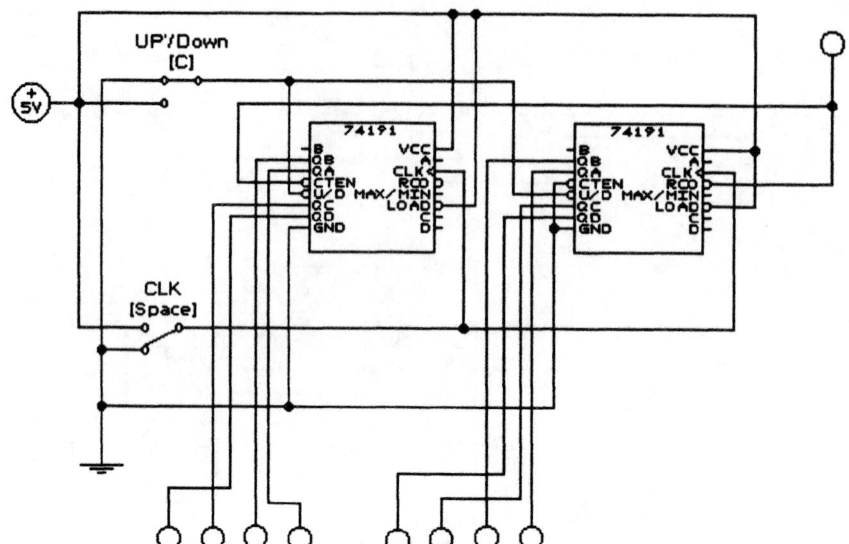

Procedure:

1. Pull down the File menu and open FIG26-1. You are looking at four negative-edge-triggered J-K flip-flops wired as a 4-bit synchronous binary counter. The C and CLK switches should be up (1). Click the On-Off switch to run the analysis and press the C key on the computer keyboard to clear the counter. Press the C key again to raise CLR' back to binary one (1).

> NOTE: If this experiment is being performed in a hardwired laboratory, use two 74112 ICs for the negative-edge-triggered J-K flip-flops and one 7408 IC for the 2-input AND gates.

Questions: Is the clear input (CLR') active low or active high?

Why does the CLR' input need to be binary one (1) in order for the counter to be able to count?

2. Press the space bar on the computer keyboard to make the CLK' input drop from one (1) to zero (0) to produce a negative clock edge. Record the counter binary output for one clock pulse in Table 26-1.

Table 26-1 Output Count

Clock Pulse	Output Q3 Q2 Q1 Q0
1	
2	
3	
4	
5	
6	
7	
8	
9	
10	
11	
12	
13	
14	
15	
16	

Question: Is the counter positive-edge-triggered or negative-edge-triggered?

3. Press the space bar on the computer keyboard enough times to produce another negative clock edge. Record the counter binary output for two clock pulses in Table 26-1. Repeat this procedure until the table is complete.

Questions: Based on the data in Table 26-1, what conclusion can you draw about the relationship between the counter binary output and the number of clock pulses applied to the counter CLK' input?

What happened when the counter was clocked after the maximum count was reached?

4. Click the On-Off switch to stop the analysis run. Pull down the File menu and open FIG26-2. You are looking at a circuit that will demonstrate the effect of flip-flop and AND gate propagation delay time (t_p) on the timing of a 4-bit synchronous binary counter. The word generator and logic analyzer settings should be as shown in Figure 26-2. Click the On-Off switch to run the analysis. The black curve plot on the logic analyzer screen is the clock (CLK') input to the counter. The blue (Q0), dark green (Q1), brown (Q2), and light green (Q3) curve plots represent the 4-bit counter output (Q3–Q0). Draw and label the CLK' input and the counter output waveshapes in the space provided.

> NOTE: If this experiment is being performed in a hardwired laboratory, use two 74112 ICs for the negative-edge-triggered J-K flip-flops and one 7408 IC for the 2-input AND gates. Use a pulse generator in place of the word generator. If a logic analyzer is not available, use a dual-trace oscilloscope.

Questions: What is the relationship between the frequencies of the flip-flop output pulses?

What is the relationship between the frequency of each flip-flop output pulse and the CLK' input pulse?

Are the output binary counts correct after each clock input pulse (negative edge)?

5. Click the On-Off switch to stop the analysis run. Change the frequency of the word generator to 100 MHz. Change the time base on the logic analyzer to 0.01 μsec/div (Version 4). Change the internal clock rate on the logic analyzer to 800 MHz (Version 5). Click the On-Off switch to run the analysis. Draw and label the new CLK' input and counter output waveshapes in the space provided.

Question: Are the output binary counts correct after each of the clock inputs (negative edges)? Explain.

6. Calculate the counter maximum clock frequency (f_{max}) based on the flip-flop propagation delay time and the AND gate delay of approximately 0.01 μsec. (See Experiment 25, Step 5, for the value of the flip-flop delay time).

Question: How does f_{max} for the synchronous counter compare with f_{max} for the asynchronous counter in Experiment 25, Step 6?

7. Click the On-Off switch to stop the analysis run. Pull down the File menu and open FIG26-3. You are looking at three negative-edge-triggered J-K flip-flops wired as a 3-bit synchronous up/down binary counter. When the UP/DOWN' switch is up, the counter will count up. When the UP/DOWN' switch is down, the counter will count down. The C, D, and CLK switches should be up (1). Click the On-Off switch to run the analysis and press the C key on the computer keyboard to clear the counter. Press the C key again to raise CLR' back to binary one (1).

NOTE: If this experiment is being performed in a hardwired laboratory, use two 74112 ICs for the negative-edge-triggered J-K flip-flops, one 7408 IC for the 2-input AND gates, one 7432 IC for the 2-input OR gates, and one 7404 IC for the INVERTER.

8. With the D switch up (count up), press the space bar on the computer keyboard to make the CLK' input drop from one (1) to zero (0) to produce a negative clock edge. Record the counter binary output for one clock pulse in Table 26-2. Press the space bar on the computer keyboard enough times to produce another negative clock edge. Record the counter binary output for two clock pulses in Table 26-2. Repeat this procedure until the table is complete.

Table 26-2 Output Count

Clock Pulse	Output Q2 Q1 Q0
1	
2	
3	
4	
5	
6	
7	
8	

Question: Based on the data in Table 26-2, what conclusion can you draw about the direction of the binary count as the clock pulses are applied to the counter CLK' input? Explain.

9. Click the On-Off switch to stop the analysis run. Press the D key on the computer keyboard to make the UP/DOWN' switch go down (count down). Click the On-Off switch to run the

analysis again and press the C key on the computer keyboard to clear the counter. Press the C key again to raise CLR' back to binary one (1).

10. With the D switch down, repeat the procedure in Step 8 and record the data in Table 26-3.

Table 26-3 Output Count

Clock Pulse	Output Q2 Q1 Q0
1	
2	
3	
4	
5	
6	
7	
8	

Question: Based on the data in Table 26-3, what conclusion can you draw about the direction of the binary count as the clock pulses are applied to the counter CLK' input? Explain.

11. Click the On-Off switch to stop the analysis run. Pull down the File menu and open FIG26-4. You will test a 74191 presettable up/down synchronous binary counter. This counter can be preset to any count by dropping the active low load input (LOAD) to binary zero (0). This will cause the binary input on the DCBA input terminals to be parallel loaded into the counter. This counter will count up one count on each positive edge of the clock pulse if the up'/down (U/D) input is set to binary zero (0). This counter will count down one count on each positive edge of the clock pulse if the up'/down (U/D) input is set to binary one (1). Notice that the UP'/DOWN switch will place a binary zero (0) on the U/D input when the switch is up, causing the counter to count up. This counter also has an active low enable input (CTEN). The active low enable input (CTEN) must be low (0) for the counter to be enabled, otherwise it will be disabled. See the Preparation section for more details about the 74191 synchronous binary counter.

12. Switches E and CLK should be down (0) and switches L and U should be up (1). Click the On-Off switch to run the analysis. Set the DCBA input to a binary number other than zero. Press the L key on the computer keyboard to lower the active low load input (LOAD) to binary zero (0). Press the L key again to raise the active low load input to binary one (1).

Question: What did you observe at the counter outputs (QD–QA)?

13. With switch U up, press the space bar on the computer keyboard to produce a positive clock edge on the CLK input. Record the outputs for one clock pulse in Table 26-4. Keep pressing the space bar to produce a number of positive clock edges on the CLK input and record the outputs in Table 26-4 until the table is complete.

Table 26-4 Output Count

Clock Pulse	Output QD QC QB QA	TC	RC
1			
2			
3			
4			
5			
6			
7			
8			
9			
10			
11			
12			
13			
14			
15			
16			

Questions: What did you observe at counter outputs QD–QA?

What did you observe at counter outputs TC and RC?

14. Press the U key on the computer keyboard to bring down the U switch. Press the space bar on the computer keyboard to produce a positive clock edge on the CLK input. Record the outputs for one clock pulse in Table 26-5. Keep pressing the space bar to produce a number of positive clock edges on the CLK input and record the outputs in Table 26-5 until the table is complete.

Table 26-5 Output Count

Clock Pulse	Output QD QC QB QA	TC	RC
1			
2			
3			
4			
5			
6			
7			
8			
9			
10			
11			
12			
13			
14			
15			
16			

Questions: What did you observe at counter outputs QD–QA?

What did you observe at counter outputs TC and RC?

15. Press the E key on the computer keyboard to raise the active low enable (CTEN) up to binary one (1). Keep pressing the space bar to apply a number of positive clock edges to the CLK input and observe the counter outputs.

Question: What did you observe at the counter outputs? Explain.

16. Click the On-Off switch to stop the analysis run. Pull down the File menu and open FIG26-5. You are looking at a circuit that will display the timing of a 74191 synchronous binary counter. The word generator and logic analyzer settings should be as shown in Figure 26-2. Switches A, B, C, and D should be down (0).

> NOTE: If this experiment is being performed in a hardwired laboratory, use a pulse generator in place of the word generator. If a logic analyzer is not available, use a dual-trace oscilloscope. Connect the load input (LOAD) to +5 V and leave inputs A, B, C, and D open. Skip Step 18.

17. Click the On-Off switch to run the analysis. The black curve plot on the logic analyzer screen is the clock input (CLK) and the first red curve plot is the parallel load input (LOAD). The blue (QA), dark green (QB), brown (QC), and light green (QD) curve plots represent the 4-bit counter output. The second red curve plot is the RCO output and the second blue curve plot is the MAX/MIN output. Draw and label the curve plots in the space provided.

Questions: What is the frequency relationship between the clock input (black) and the counter outputs QA (blue), QB (dark green), QC (brown), and QD (light green)? What is the modulus (divide-by) of this counter?

Is the counter counting up or down? Explain why.

18. Click the On-Off switch to stop the analysis run. Change switches A, B, C, and D to a binary number other than zero (0000). This will cause the counter to start counting at a binary number other than zero. Click the On-Off switch to run the analysis again. Draw and label the new curve plots in the space provided.

Questions: What is the first binary output (QD–QA)? Is it the same as the binary input DCBA?

Is the counter counting up or down?

At what output count does output RCO go low (0)?

At what output count does output MAX/MIN go high (1)?

19. Click the On-Off switch to stop the analysis run. Pull down the File menu and open FIG26-6. You will test a 74191 wired as a MOD-12 (divide-by-12) counter. Notice that the MAX/MIN output and the CLK input are connected to a NAND gate that feeds the LOAD input. This will cause the counter to reset the count during the positive half of the clock period when the counter output reaches the terminal state and the MAX/MIN output is high (1). The counter will reset to the binary count that is set by the DCBA switches.

20. Click the On-Off switch to run the analysis. Press the space bar on the computer keyboard enough times to apply a series of positive clock edges to the CLK input and observe the output (QD–QA). Record the results in Table 26-6.

Table 26-6 Output Count

Clock Pulse	Output QD QC QB QA
1	
2	
3	
4	
5	
6	
7	
8	
9	
10	
11	
12	
13	
14	
15	
16	

Questions: Is the counter counting up or down?

What is the number of different binary output states for this counter?

What happened on clock pulse number twelve? Explain.

Based on these results, what is the modulus (divide-by) of this counter?

How does the modulus (MOD) of this counter compare with the parallel input (DCBA) binary number?

How can you change the modulus (MOD) of this counter without changing the wiring?

21. Click the On-Off switch to stop the analysis run. Change the modulus (divide-by) of this counter to MOD-10 without changing the wiring. Repeat Step 20. Record the results in Table 26-7.

Table 26-7 Output Count

Clock Pulse	Output QD QC QB QA
1	
2	
3	
4	
5	
6	
7	
8	
9	
10	
11	
12	
13	
14	
15	
16	

Questions: What is the number of different binary output states for this counter?

Based on these results, what is the modulus (divide-by) of this counter? Is it as expected?

22. Click the On-Off switch to stop the analysis run. Pull down the File menu and open FIG26-7. The function generator and oscilloscope settings should be as shown in Figure 26-7. Click the On-Off switch to run the analysis. Use the "zoom" feature on the oscilloscope to determine the time period (T) for one cycle and the frequency (f) of the clock input (black) and the counter output (blue). Record your answers in the space provided.

Clock input T_c = ____________

f_c = ____________

Counter output T_o = ____________

f_o = ____________

23. Based on the results in Step 22, calculate the modulus (divide-by) of this counter.

Questions: Based on the circuit configuration in Figure 26-7, what is the expected modulus (divide-by) of this counter? Explain.

How does this answer compare with the calculated modulus in Step 23?

24. Click the On-Off switch to stop the analysis run. Change the modulus (divide-by) of the counter to MOD-8 without changing the wiring. Click the On-Off switch to run the analysis. Use the "zoom" feature on the oscilloscope to determine the time period (T) for one cycle and the frequency (f) of the clock input (black) and the counter output (blue). Record your answers in the space provided.

Clock input T_c = ____________

f_c = ____________

Counter output T_o = ____________

f_o = ____________

25. Based on the results in Step 24, calculate the modulus (divide-by) of this counter.

Question: How does this answer in Step 25 compare with the expected modulus (divide-by)?

26. Click the On-Off switch to stop the analysis run. Pull down the File menu and open FIG26-8. You are looking at two 74191 cascaded synchronous counters wired as an 8-bit counter. The counter on the right outputs the low 4-bits and the counter on the left outputs the high 4-bits. The cascaded 8-bit synchronous counter can count up or down depending on the position of the UP'/DOWN switch. Notice that the RCO output of the counter on the right is connected to the active low enable input (CTEN) of the counter on the left. This will cause the counter on the left to be enabled during the low half of the clock period and increment or decrement one count when the counter on the right reaches its terminal count and begins a new count cycle.

27. Switch C should be up and the CLK switch should be down (0). Click the On-Off switch to run the analysis. Press the space bar on the computer keyboard enough times to apply a series of positive clock edges to the CLK input to make the counter count the full range of the count and observe the output.

Questions: Is the counter counting up or down?

What is the maximum count of this counter in binary and decimal?

What is the modulus (MOD) of this counter?

28. Click the On-Off switch to stop the analysis run. Press the C key on the computer keyboard to lower the UP'/DOWN switch. Make sure the CLK switch is up (1). Click the On-Off switch to run the analysis. Press the space bar on the computer keyboard enough times to apply a series of positive clock edges to the CLK input to make the counter count the full range of the count and observe the output.

Question: Is the counter counting up or down? Explain.

Name ______________________________
Date ______________________________

BCD Counters

Objectives:

1. Demonstrate the application of J-K flip-flops to asynchronous BCD counters.
2. Demonstrate the operation of a 7490 asynchronous decade counter as a BCD counter.
3. Investigate 7490 asynchronous BCD counter timing.
4. Demonstrate the application of J-K flip-flops to synchronous BCD counters.
5. Demonstrate the operation of the 74190 presettable synchronous decade counter as a BCD counter.
6. Investigate 74190 synchronous BCD counter timing.
7. Investigate cascaded BCD counters.
8. Build a BCD counter display system using a 7 segment display.

Materials:

One 5 V dc power supply
Eight logic switches
Ten logic probe lights
Four negative-edge-triggered J-K flip-flops (2-74112 ICs)
One asynchronous decade counter (1-7490 IC)
Two presettable synchronous decade counters (2-74190 ICs)
Four two-input AND gates (1-7408 IC)
One two-input OR gate (1-7432 IC)
One INVERTER (1-7404 IC)
Two BCD-to-7 segment decoder drivers (2-7448 ICs)
Two common cathode 7 segment LED displays
One function (pulse) generator
One logic analyzer or dual-trace oscilloscope

Preparation:

Make sure you complete Experiments 25 and 26 before attempting this experiment. Also, review the Preparation sections of those experiments.

Counters that have ten output states (MOD-10) are often referred to as decade counters, regardless of the sequence of the count. Decade counters are often used for dividing a pulse frequency by ten. A decade counter with a count sequence of zero (0000) through nine (1001) is commonly called a BCD counter because its ten output states consist of the BCD code. BCD counters find widespread use in applications where pulses or events are to be counted with the results displayed on a decimal readout.

In order to build a decade or BCD counter, it is necessary to force a 4-bit counter to reset (clear) before completing all of its normal output states. A BCD counter must reset (clear) to binary zero (0000) on the tenth count after reaching binary nine (1001). Therefore, it will count between zero (0000) and nine (1001). One way to make a binary counter reset (clear) after the count of nine (1001) is to decode count ten (1010) with a NAND gate and use the output of the NAND gate to reset (clear) the counter. This will cause a glitch on the Q1 output just before the counter resets (clears). This glitch can be eliminated by using logic circuitry to control the timing of each counter flip-flop instead of using the reset (clear) counter input to clear the counter. This requires a more complex circuit and raises the cost of the counter IC chip.

The circuit in Figure 27-1 consists of four negative-edge-triggered J-K flip-flops wired as an asynchronous BCD counter using logic circuitry to control the timing of each counter flip-flop. Because the flip-flop J and K inputs are controlled by different logical outputs to force each flip-flop to toggle at the correct time for a zero (0000) to nine (1001) count, this counter will not produce any glitches in the output pulses. This circuit is similar to the circuit in the popular 7490 asynchronous decade counter, except the output of the first flip-flop (Q0 or QA) is not connected internally to the clock input of the second flip-flop in the 7490. This counter is asynchronous because all of the flip-flops are not clocked simultaneously. The counter active low CLR' input is formed by connecting all of the flip-flop active low dc clear inputs together. When the CLR' input is dropped to binary zero (0), all of the flip-flops will clear (0), causing the counter to be cleared. The counter CLR' input must be returned to binary one (1) in order for the counter to count. If the active low CLR' input is not returned to binary one (1), the counter will stay cleared because the flip-flop dc clear inputs override the flip-flop CLK inputs.

The 7490 in Figure 27-2 is an asynchronous decade counter. It is wired internally as a 1-bit MOD-2 (divide-by-2) counter with a negative-edge-triggered clock input (CLKA) and output QA, and a 3-bit MOD-5 (divide-by-5) counter with a negative-edge-triggered clock input (CLKB) and outputs QB, QC, and QD. The 7490 in Figure 27-2 is externally wired as a BCD counter by connecting the output of the 1-bit MOD-2 counter (QA) to the clock input of the 3-bit MOD-5 counter (CLKB). This feature makes the 7490 very versatile because it can be wired as a 1-bit (MOD-2), 3-bit (MOD-5), or decade (MOD-10) counter. The 7490 also has two active high reset (clear) inputs (RO1 and RO2) and two active high set inputs (R91 and R92). The reset (clear) and set inputs can be used to reset the counter or cut off the binary count in order to change the modulus (divide-by) of the counter. By wiring the correct combination of counter outputs to the reset (clear) inputs (RO1 and RO2) or the set inputs (R91 and R92), the modulus (divide-by) of the counter can be changed to any value between MOD-2 (divide-by-2) and MOD-10 (divide-by-10).

The circuit in Figure 27-3 will display the timing of the 7490 asynchronous decade counter configured as a BCD counter. The word generator will apply the clock pulses to the counter clock input (CLKA). The logic analyzer will monitor the clock pulses and the counter outputs (QA–QD).

The active high reset (clear) inputs (RO1 and RO2) and the active high set inputs (R91 and R92) are permanently connected to ground (0) to allow the counter to count.

The circuit in Figure 27-4 consists of four negative-edge-triggered J-K flip-flops wired as a synchronous BCD counter. This is a synchronous counter because all of the flip-flop clock inputs are connected to the counter CLK' input and are clocked simultaneously. Each flip-flop J and K input is connected to either 5 V (J=1, K=1) or the output of a logic network. The correct logic causes the flip-flops to toggle at the proper time to produce a zero (0000) to nine (1001) count. The counter active low CLR' input is formed by connecting all of the flip-flop active low dc clear inputs together. When the CLR' input is dropped to binary zero (0), all of the flip-flops will clear (0), causing the counter to be cleared. The counter CLR' input must be returned to binary one (1) in order for the counter to count. Because this synchronous counter has more extensive circuitry than the asynchronous counter, it is more expensive to build but it can be clocked at a higher clock frequency.

The 74190 in Figure 27-5 is a presettable synchronous decade counter. It can be preset to any count by applying a 4-bit binary number to inputs DCBA and dropping the active low load input (LOAD) to binary zero (0). The load input (LOAD) must be returned to binary one (1) in order for the counter to count because the load input overrides the CLK input. The 74190 will count up if a binary zero (0) is applied to the U/D input and it will count down if a binary one (1) is applied to the U/D input. The counter will increment or decrement one count each time the CLK input receives a positive clock edge. The active low count enable input (CTEN) must be low (0) for the counter to be enabled, otherwise it will be disabled. The TC (MAX/MIN) output is normally low (0), but will go high (1) for one clock period when the counter reaches zero (0000) in the count down mode or nine (1001) in the count up mode. The RC (RCO) output is normally high (1), but will go low (0) for the low half of the clock period when the counter reaches zero (0000) in the count down mode or nine (1001) in the count up mode. The TC (MAX/MIN) and RC (RCO) outputs are made available in order to be able to cascade 74190 synchronous decade counters to provide a count higher than BCD nine (1001). Each additional cascaded counter adds a decimal digit to the output.

The circuit in Figure 27-6 will display the timing of the 74190 synchronous decade counter. The word generator will apply the clock pulses to the counter clock input. The logic analyzer will monitor the clock pulses, the counter outputs (QA–QD), the ripple clock output (RCO), and the terminal count output (MAX/MIN).

The circuit in Figure 27-7 consists of two 74190 cascaded synchronous BCD counters. The counter on the right outputs the least significant BCD digit and the counter on the left outputs the most significant BCD digit. This cascaded synchronous BCD counter can count up or down depending on the position of the UP'/DOWN switch. Notice that the RCO output of the counter on the right is connected to the active low enable input (CTEN) of the counter on the left. This will cause the counter on the left to be enabled during the low half of the clock period and increment or decrement one count when the counter on the right reaches its terminal count and begins a new count cycle. Because both counters are being clocked simultaneously, this cascaded pair is synchronous.

The circuit in Figure 27-8 consists of two cascaded 74190 synchronous BCD counters connected to two BCD-to-7 segment decoder/drivers driving LED displays. The 7 segment LED displays will output the decimal equivalent of the BCD output code from each of the BCD counters. The right display will output the least significant decimal digit and the left display will output the most significant decimal digit. The CLK logic switch will provide a positive clock edge to clock both synchronous BCD counters simultaneously, making the display system synchronous. Grounding the ripple blanking input (RBI) on the left BCD-to-7 segment decoder will cause the left display to be blank when the BCD input is zero (0000).

Figure 27-1 Asynchronous BCD Counter

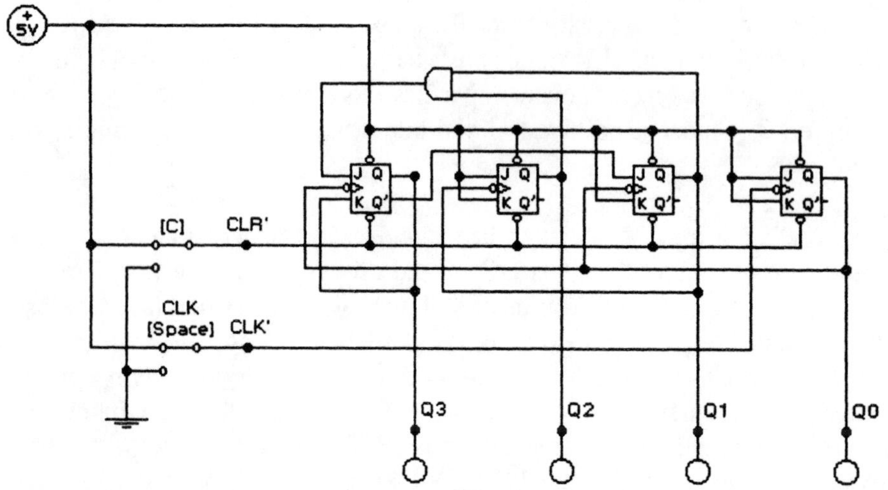

Figure 27-2 7490 Asynchronous BCD Counter

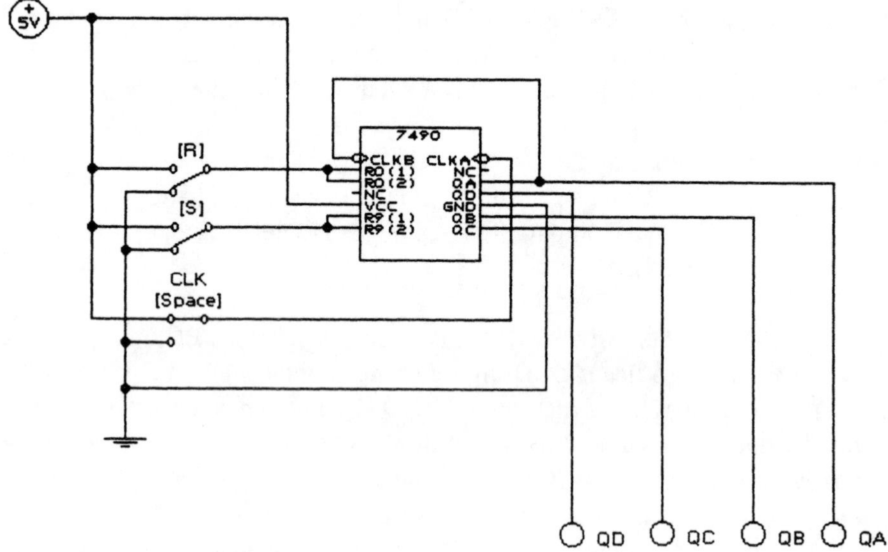

Figure 27-3a 7490 Asynchronous BCD Counter Timing (EWB Version 4)

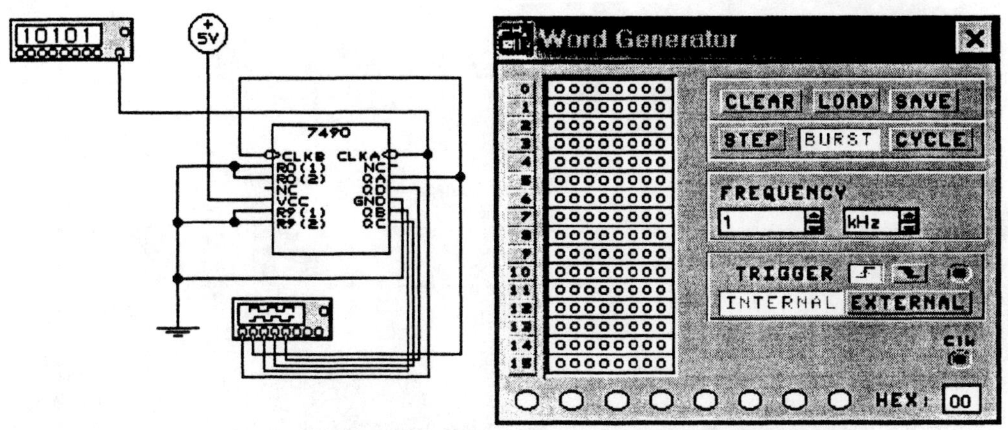

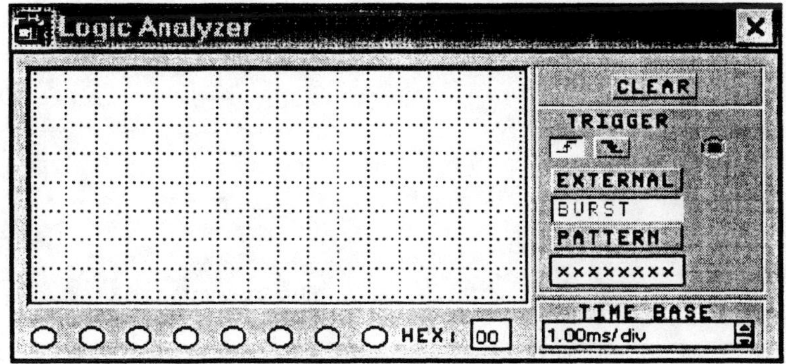

Figure 27-3b 7490 Asynchronous BCD Counter Timing (EWB Version 5)

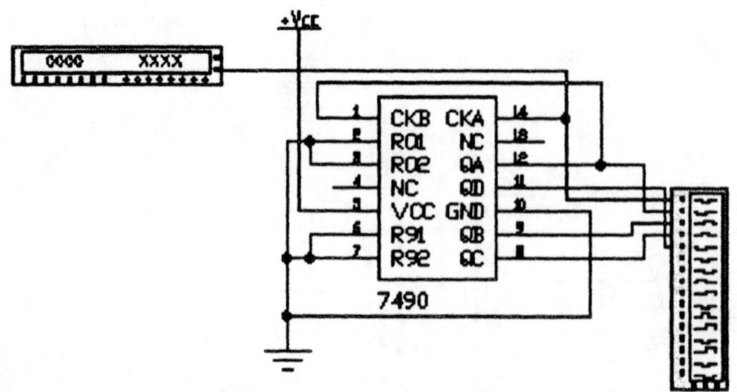

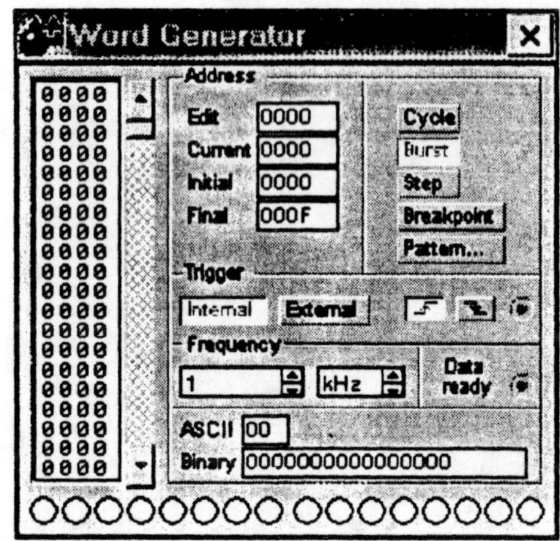

Logic Analyzer Settings
Clocks per division ----------------8

Clock Setup dialog box
Clock edge --------------- positive
Clock mode -------------- Internal
Internal clock rate ------ 10 kHz
Clock qualifier ----------- x
Pre-trigger samples --- 100
Post-trigger samples -- 1000
Threshold voltage (V) 2

Trigger Patterns dialog box
A ------------------------------ xxxxxxxxxxxxxxxx
Trigger combinations -- A
Trigger qualifier ---------- x

Figure 27-4 Synchronous BCD Counter

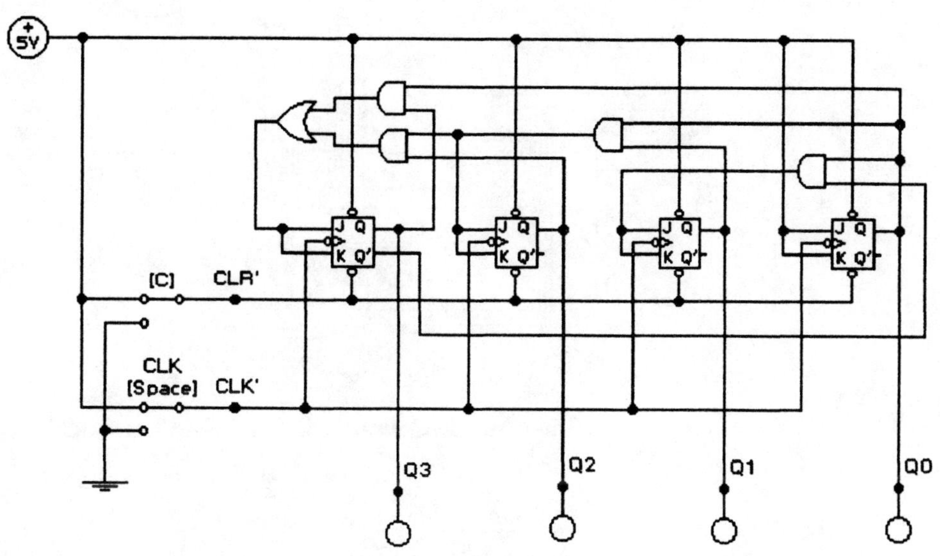

Figure 27-5 74190 Presettable Synchronous Decade Counter

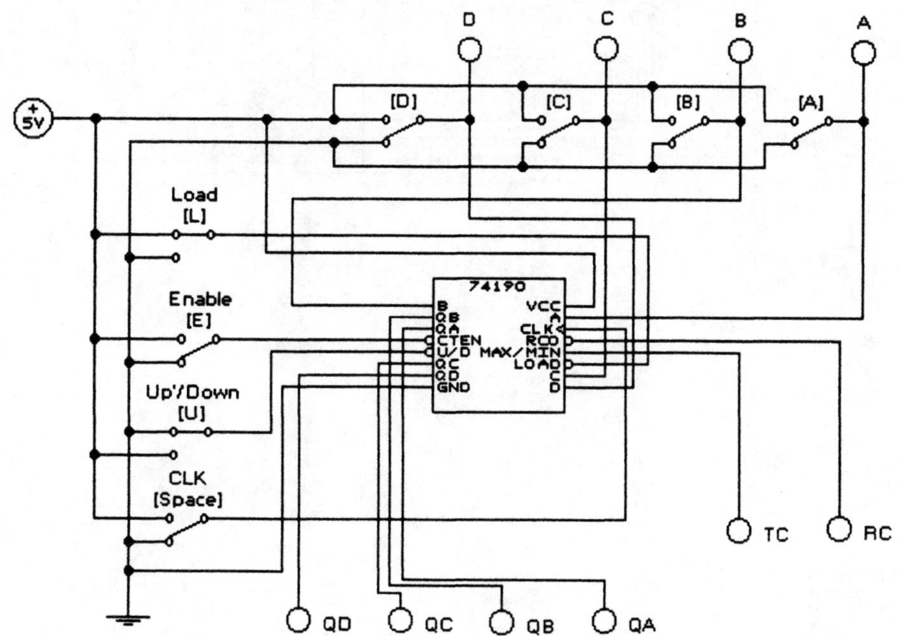

Figure 27-6a 74190 Synchronous Decade Counter Timing (EWB Version 4)

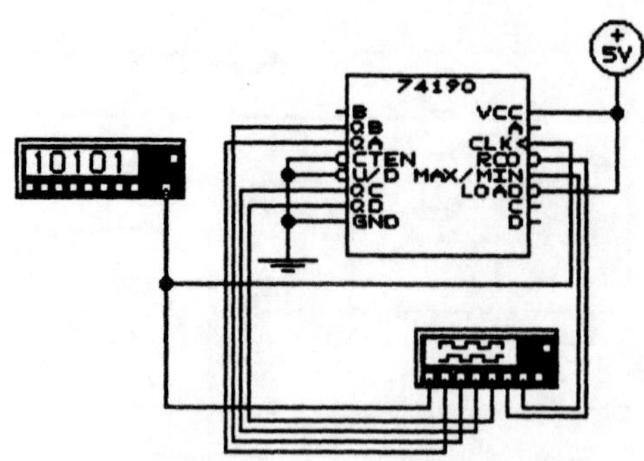

Figure 27-6b 74190 Synchronous Decade Counter Timing (EWB Version 5)

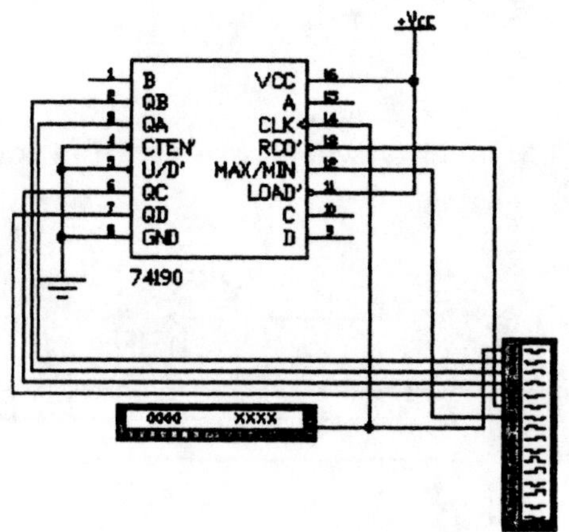

Figure 27-7 Cascaded 74190 Synchronous BCD Counters

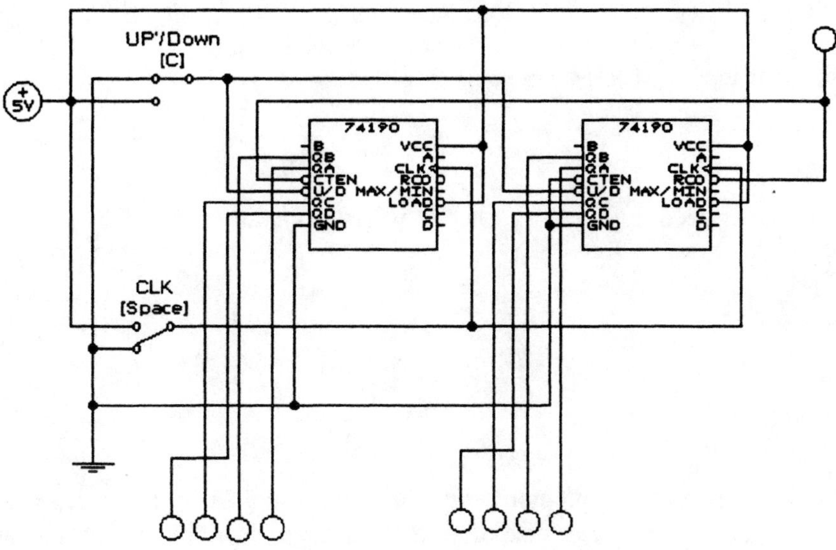

Figure 27-8 BCD Counter Display System

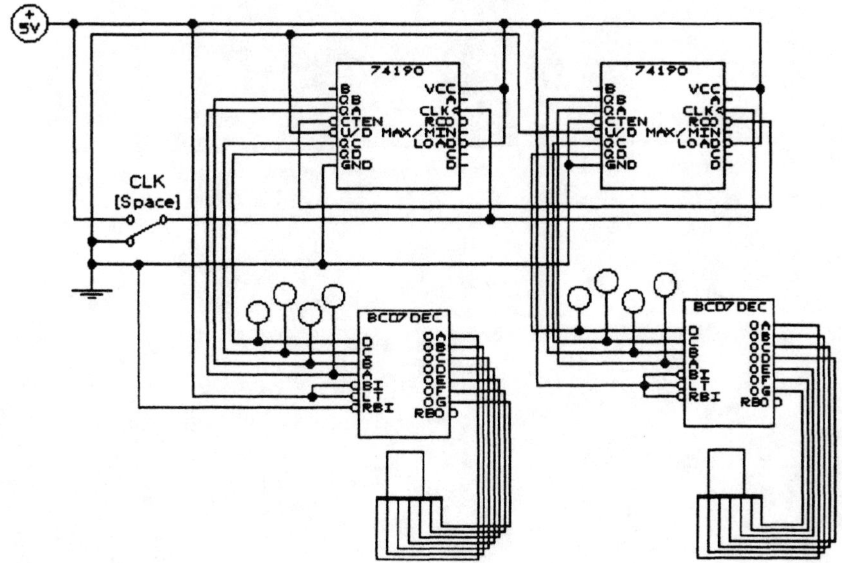

Procedure:

1. Pull down the File menu and open FIG27-1. You are looking at four negative-edge-triggered J-K flip-flops wired as an asynchronous BCD counter. The C and CLK switches should be up (1). Click the On-Off switch to run the analysis and press the C key on the computer keyboard to clear the counter. Press the C key again to raise CLR' back to binary one (1).

NOTE: If this experiment is being performed in a hardwired laboratory, use two 74112 ICs for the negative-edge-triggered J-K flip-flops and one 7408 IC for the 2-input AND gate.

Questions: Is the clear input (CLR') active low or active high?

Why does the CLR' input need to be binary one (1) in order for the counter to be able to count?

2. Press the space bar on the computer keyboard to make the CLK' input drop from one (1) to zero (0) to produce a negative clock edge. Record the counter binary output for one clock pulse in Table 27-1.

Table 27-1 Output Count

Clock Pulse	Output Q3 Q2 Q1 Q0
1	
2	
3	
4	
5	
6	
7	
8	
9	
10	
11	
12	
13	
14	
15	
16	

Question: Is the counter positive-edge-triggered or negative-edge-triggered?

3. Press the space bar on the computer keyboard enough times to produce another negative clock edge. Record the counter binary output for two clock pulses in Table 27-1. Repeat this procedure until the table is complete.

Questions: Based on the data in Table 27-1, what conclusion can you draw about the relationship between the counter binary output and the number of clock pulses applied to the counter CLK' input?

What was the maximum count?

What happened when the counter was clocked after the maximum count was reached?

4. Click the On-Off switch to stop the analysis run. Pull down the File menu and open FIG27-2. You will test a 7490 asynchronous decade counter. This counter has two active high reset (clear) inputs (RO1 and RO2), two active high set inputs (R91 and R92), and two negative-edge-triggered clock inputs (CLKA and CLKB). The CLKA input is used to trigger flip-flop A and the CLKB input is used to trigger the 3-bit MOD-5 counter using flip-flops B, C, and D. The 7490 in Figure 27-2 is externally wired as a BCD counter by connecting flip-flop output QA to clock input CLKB and using CLKA as the clock input. See the Preparation section for more details about the 7490 asynchronous decade counter.

5. Switches R and S should be down (0) and the CLK switch should be up (1). Click the On-Off switch to run the analysis. If the counter output (QD–QA) is not cleared, press the R key on the computer keyboard to raise the reset (clear) input (RO1 and RO2) to binary one (1) to clear the counter. Press the R key again to lower the reset input back to binary zero (0).

6. Press the space bar on the computer keyboard to produce a negative clock edge on the CLKA input. Keep pressing the space bar to produce a number of negative clock edges on the CLKA input and cycle the counter through its full count range.

Question: What did you observe at counter outputs QD–QA?

7. Press the S key on the computer keyboard and observe the counter output (QD–QA).

Question: What did you observe at counter outputs QD–QA?

8. Click the On-Off switch to stop the analysis run. Pull down the File menu and open FIG27-3. You are looking at a circuit that will display the timing of a 7490 asynchronous decade counter configured as a BCD counter, using CLKA as the clock input. The word generator and logic analyzer settings should be as shown in Figure 27-3.

NOTE: If this experiment is being performed in a hardwired laboratory, use a pulse generator in place of the word generator. If a logic analyzer is not available, use a dual-trace oscilloscope.

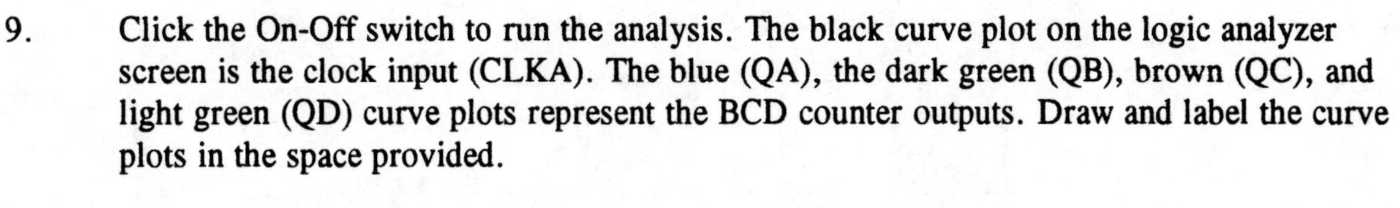

9. Click the On-Off switch to run the analysis. The black curve plot on the logic analyzer screen is the clock input (CLKA). The blue (QA), the dark green (QB), brown (QC), and light green (QD) curve plots represent the BCD counter outputs. Draw and label the curve plots in the space provided.

Questions: Are the output binary counts correct after each pulse (negative edge)?

What happened on the tenth clock pulse (negative edge)?

What is the frequency relationship between the CLKA input (black) and counter output QD (light green)? What is the modulus (divide-by) of this counter?

10. Click the On-Off switch to stop the analysis run. Pull down the File menu and open FIG27-4. You are looking at four negative-edge-triggered J-K flip-flops wired as a synchronous BCD counter. The C and CLK switches should be up (1). Click the On-Off switch to run the analysis and press the C key on the computer keyboard to clear the counter. Press the C key again to raise CLR' back to binary one (1).

NOTE: If this experiment is being performed in a hardwired laboratory, use two 74112 ICs for the negative-edge-triggered J-K flip-flops, one 7408 IC for the 2-input AND gates, and one 7432 IC for the 2-input OR gate.

11. Press the space bar on the computer keyboard to make the CLK' input drop from one (1) to zero (0) to produce a negative clock edge. Record the counter binary output for one clock pulse in Table 27-2.

Table 27-2 Output Count

Clock Pulse	Output Q3 Q2 Q1 Q0
1	
2	
3	
4	
5	
6	
7	
8	
9	
10	
11	
12	
13	
14	
15	
16	

Question: Is the counter positive-edge-triggered or negative-edge-triggered?

12. Press the space bar on the computer keyboard enough times to produce another negative clock edge. Record the counter binary output for two clock pulses in Table 27-2. Repeat this procedure until the table is complete.

Questions: Based on the data in Table 27-2, what conclusion can you draw about the relationship between the counter binary output and the number of clock pulses applied to the counter CLK' input?

What was the maximum count?

What happened when the counter was clocked after the maximum count was reached?

How did these outputs compare with the outputs of the asynchronous counter in Table 27-1?

13. Click the On-Off switch to stop the analysis run. Pull down the File menu and open FIG27-5. You will test a 74190 presettable synchronous decade counter. This counter can be preset to any count by dropping the active low LOAD input to binary zero (0). This will cause the binary input on the DCBA input terminals to be parallel loaded into the counter. This counter will count up one count on each positive edge of the clock pulse if binary zero (0) is applied to the up'/down (U/D) input, and count down one count on each positive edge of the clock pulse if binary one (1) is applied to the up'/down (U/D) input. Notice that the UP'/DOWN switch will place a binary zero (0) on the U/D input when the switch is up, causing the counter to count up. This counter also has an active low enable input (CTEN), which must be low (0) for the counter to be enabled, otherwise it will be disabled. See the Preparation section for more details about the 74190 synchronous decade counter.

14. Switches E and CLK should be down (0) and switches L and U should be up (1). Click the On-Off switch to run the analysis. Make sure the DCBA input is set to binary zero (0000). If the counter is not cleared, press the L key on the computer keyboard to lower the active low LOAD input to binary zero (0) and load all zeros into the counter. Press the L key again to raise the active low LOAD input to binary one (1).

15. With switch U up, press the space bar on the computer keyboard to produce a positive clock edge on the CLK input. Record the outputs for one clock pulse in Table 27-3. Keep pressing the space bar to produce a number of positive clock edges on the CLK input and record the outputs in Table 27-3 until the table is complete.

Table 27-3 Output Count

Clock Pulse	Output QD QC QB QA	TC	RC
1			
2			
3			
4			
5			
6			
7			
8			
9			
10			
11			
12			
13			
14			
15			
16			

Questions: What did you observe at counter outputs QD–QA?

What was the maximum count?

What happened when the counter was clocked after the maximum count was reached?

What did you observe at counter outputs TC and RC?

16. Set the DCBA switches to a binary nine (1001). Press the L key to load the counter with nine (1001). Press the L key again to raise the LOAD input back to binary one (1). Press the U key to bring down the U switch. Press the space bar on the computer keyboard to produce a positive clock edge on the CLK input. Record the outputs for one clock pulse in Table 27-4. Keep pressing the space bar to produce a number of positive clock edges on the CLK input and record the outputs in Table 27-4 until the table is complete.

Table 27-4 Output Count

Clock Pulse	Output QD QC QB QA	TC	RC
1			
2			
3			
4			
5			
6			
7			
8			
9			
10			
11			
12			
13			
14			
15			
16			

Questions: What did you observe at counter outputs QD–QA?

What did you observe at counter outputs TC and RC?

17. Press the E key on the computer keyboard to raise the active low enable (CTEN) to binary one (1). Keep pressing the space bar to apply a number of positive clock edges to the CLK input and observe the counter outputs.

Question: What did you observe at the counter outputs? Explain.

18. Click the On-Off switch to stop the analysis run. Pull down the File menu and open FIG27-6. You are looking at a circuit that will display the timing of a 74190 synchronous decade counter. The word generator and logic analyzer settings should be as shown in Figure 27-3.

NOTE: If this experiment is being performed in a hardwired laboratory, use a pulse generator in place of the word generator. If a logic analyzer is not available, use a dual-trace oscilloscope.

19. Click the On-Off switch to run the analysis. The black curve plot on the logic analyzer screen is the clock input (CLK). The blue (QA), dark green (QB), brown (QC), and light green (QD) curve plots represent the counter outputs. The red curve plot is the RCO output and the second blue curve plot is the MAX/MIN output. Draw and label the curve plots in the space provided.

Questions: Is the counter counting up or down? Explain why.

Are the output binary counts correct after each clock pulse (positive edge)?

What did you observe at outputs RCO and MAX/MIN?

What is the frequency relationship between the clock input (black) and counter output QD (light green)? What is the modulus (divide-by) of this counter?

20. Click the On-Off switch to stop the analysis run. Pull down the File menu and open FIG27-7. You are looking at two cascaded 74190 synchronous BCD counters. The counter on the right outputs the least significant BCD digit and the counter on the left outputs the most significant BCD digit. This cascaded synchronous BCD counter can count up or down depending on the position of the UP'/DOWN switch. Notice that the RCO output of the counter on the right is connected to the active low enable input (CTEN) of the counter on the left. This will cause the counter on the left to be enabled during the low half of the clock period and increment or decrement one count when the counter on the right reaches its terminal count and begins a new count cycle.

21. Switch C should be up and the CLK switch should be down (0). Click the On-Off switch to run the analysis. Press the space bar on the computer keyboard enough times to apply a series of positive clock edges to the CLK input to make the counter count the full range of the count and observe the output.

Questions: Is the counter counting in BCD or binary?

Is the counter counting up or down?

What is the maximum count for each counter?

What is the maximum count for this cascaded counter?

22. Click the On-Off switch to stop the analysis run. Press the C key on the computer keyboard to lower the UP'/DOWN switch. Make sure the CLK switch is up (1). Click the On-Off switch to run the analysis. Press the space bar on the computer keyboard enough times to apply a series of positive clock edges to the CLK input to make the counter count the full range of the count and observe the output.

Questions: Is the counter counting in BCD or binary?

Is the counter counting up or down? Explain why.

23. Click the On-Off switch to stop the analysis run. Pull down the File menu and open FIG27-8. You are looking at two cascaded 74190 synchronous BCD counters connected to two BCD-to-7 segment decoder/drivers driving two LED displays. The 7 segment displays will output the decimal equivalent of the BCD output code for each of the BCD counters. The clock switch should be down (0). See the Preparation section for more details.

NOTE: It is recommended that this step be performed in a hardwired laboratory to obtain experience wiring and testing an actual logic circuit. Use two 7448 BCD-to-7 segment decoder/drivers and two common cathode LED displays.

24. Click the On-Off switch to run the analysis. Press the space bar on the computer keyboard enough times to apply a series of positive clock edges to the CLK input to make the counter count the full range of the count and observe the output.

Questions: What was the highest count for each BCD counter?

What happened after the counter on the right reached its highest count?

What was the highest count of the cascaded pair?

What would be required to produce a higher count?

Did the BCD output from each BCD counter match the decimal output on the LED displays?

What caused the left LED display to be blanked when the BCD input was zero (0000)?

Name ____________________

Date ____________________

EXPERIMENT

28 Troubleshooting Sequential Logic Circuits

Objectives:

1. Determine the defective flip-flop in a 4-bit asynchronous (ripple) counter wired using four J-K flip-flops.
2. Determine the defective component in a 4-bit synchronous counter wired using four J-K flip-flops and two 2-input AND gates.
3. Determine the defective counter in a cascaded counter pair.
4. Determine the correct output frequency for a 74191 presettable synchronous counter wired as a frequency divider.
5. Determine if the output waveshapes are correct for a particular 74191 presettable synchronous counter circuit configuration.

Materials:

This experiment can only be performed on Electronics Workbench using the circuits disk provided with this manual.

Preparation:

In order to perform these experiments effectively, you must first complete Experiments 21–27. Also review the Preparation sections of those experiments before beginning these troubleshooting exercises. Use the theory learned in Experiments 21–27 to answer the questions or find the defective component in the logic circuits in these experiments.

Procedure:

1. Pull down the File menu and open FIG28-1. You are looking at a 4-bit asynchronous (ripple) counter wired using four J-K flip-flops. Click the On-Off switch to run the analysis. Double-click the logic analyzer to bring down the enlargement. Based on the waveshapes on the logic analyzer screen, determine which flip-flop (Q0, Q1, Q2, or Q3) is defective. Explain how you determined your answer.

   Defective flip-flop _______

2. Pull down the File menu and open FIG28-2. You are looking at a 4-bit synchronous counter wired using four J-K flip-flops and two 2-input AND gates. Click the On-Off switch to run the analysis. Double-click the logic analyzer to bring down the enlargement. Based on the waveshapes on the logic analyzer screen, determine which component (flip-flops Q0, Q1, Q2, Q3 or AND gates G1 or G2) is defective. Explain how you determined your answer.

   Defective component _______

3. Pull down the File menu and open FIG28-3. You are looking at a 4-bit synchronous counter wired using four J-K flip-flops and two 2-input AND gates. Click the On-Off switch to run the analysis. Double-click the logic analyzer to bring down the enlargement. Based on the waveshapes on the logic analyzer screen, determine which component (flip-flops Q0, Q1, Q2, Q3 or AND gates G1 or G2) is defective. Explain how you determined your answer.

Defective component ________

4. Pull down the File menu and open FIG28-4. You are looking at a 4-bit synchronous counter wired using four J-K flip-flops and two 2-input AND gates. Click the On-Off switch to run the analysis. Double-click the logic analyzer to bring down the enlargement. Based on the waveshapes on the logic analyzer screen, determine which component (flip-flops Q0, Q1, Q2, Q3 or AND gates G1 or G2) is defective. Explain how you determined your answer.

Defective component ________

5. Pull down the File menu and open FIG28-5. You are looking at cascaded counters. Click the On-Off switch to run the analysis. Use the zoom feature on the oscilloscope screen to determine the output time period (T) and the output frequency (f) for each counter. Record your answers in the space provided.

T_1 = ______________ f_1 = ______________

T_2 = ______________ f_2 = ______________

Question: Based on the wiring configuration of each counter and the input frequency of 60 kHz, are the output frequencies correct? If not, which counter is defective?

6. Pull down the File menu and open FIG28-6. Click the On-Off switch to run the analysis. Use the zoom feature on the oscilloscope screen to determine the output (blue curve plot) time period (T) and the output frequency (f) for the counter. Record your answers in the space provided.

T = ______________ f = ______________

Question: Based on the circuit configuration in Figure 28-6, is the 74191 counter producing the correct output frequency? If not, what should the frequency be?

7. Pull down the File menu and open FIG28-7. Click the On-Off switch to run the analysis.

Question: Are the output waveshapes correct for the 74191 presettable synchronous counter circuit configuration in Figure 28-7? If not, what is wrong with the output waveshapes?

Part

Interfacing the Analog World

In the experiments in Part V, you will learn how the digital world is interfaced with the analog world. Most real world physical variables are analog in nature and have values that are within a continuous range. On the other hand, digital systems are discrete in nature and have values that are one of two possibilities. Any analog data that must be input into a digital system must first be converted into digital form. Any digital data that must be output into the real analog world must first be converted into analog form. A circuit that converts digital codes into analog data is called a digital-to-analog converter (DAC). A circuit that converts analog data into digital codes is called an analog-to-digital converter (ADC). Digital-to-analog converters and analog-to-digital converters will be studied in the first two experiments in Part V.

There are many applications in which analog data must be converted into digital codes and transferred into a digital system. This process is called data acquisition. The principles of data acquisition will be studied in the last experiment in Part V.

> If the experiments in Part V are performed in a hardwired laboratory, make sure you save the D/A converter circuit wired in Experiment 29 and the A/D converter circuit wired in Experiment 30. They will be used in the data acquisition experiment (Experiment 31).

The circuits for the experiments in Part V can be found on the enclosed disk in the PART5 subdirectory.

Name ____________________
Date ____________________

Digital-to-Analog Converters

Objectives:

1. Demonstrate the relationship between the digital input and the analog output of a D/A converter (DAC).
2. Demonstrate how to set the full-scale output (range) of a D/A converter.
3. Demonstrate how to measure the output offset of a D/A converter.
4. Develop an understanding of the concept of resolution as applied to D/A converters.
5. Demonstrate how a staircase output can be used to test a D/A converter and measure its resolution.

Materials:

One dc voltage supply (+5 V, +12 V, -12 V)
Eight logic switches
Eight logic probe lights
One D/A converter (1-1408 or DAC0808 IC)
One op-amp (1-741 IC)
One asynchronous counter (1-7493 IC)
Two synchronous counters (2-74191 ICs)
One INVERTER (1-7404 IC)
One 0–2 kΩ potentiometer
One 0.1 μF capacitor
Two 1 kΩ resistors
One 0–10 V voltmeter
One function (pulse) generator
One oscilloscope

Preparation:

Most real world physical variables are analog in nature and have values that are within a continuous range. On the other hand, digital systems are discrete in nature and have values that are one of two possibilities. Any digital data that must be output into the real analog world must first be converted into analog form. A circuit that converts digital codes into analog data is called a digital-to-analog converter (DAC). The output of a D/A converter (DAC) can be in the form of a voltage or a current.

A D/A converter (DAC) will convert a digital binary input to an output voltage or current that is proportional to the magnitude of the binary input value. The DAC full-scale output is the output value when all ones are applied to the DAC binary input. The full-scale output value determines the range of the DAC.

The DAC output offset is the value of the output when all binary zeros are applied to the DAC binary input. In an ideal DAC, the output offset is zero. In a real DAC, the output offset is not zero. Many DACs have an external offset adjustment that allows you to zero the output offset.

The resolution of a DAC is defined as the smallest change that can occur in the analog output as a result of a change in the digital input. The DAC resolution (step size) can be measured by measuring the change in the output voltage or current for a single step change in the binary input. The calculated DAC resolution (step size) is equal to the full-scale output divided by the number of input steps from zero to full-scale output. The number of input steps depends solely on the number of DAC input bits, where a 4-bit DAC has 15 steps and an 8-bit DAC has 255 steps (0 is not a step). The percent resolution is equal to the percentage of the full-scale output that one step represents. Therefore, the percentage resolution can be calculated by dividing the step size by the full-scale output times 100%, which is equal to the inverse of the number of steps times 100%. This means that the larger the number of bits, the higher the number of steps and the smaller the resolution (step size). For this reason, most manufacturers usually specify a DAC resolution as the number of bits.

The 8-bit voltage output DAC circuit in Figure 29-1a will help develop an understanding of the relationship between the digital input and the analog output of a DAC. The DAC full-scale output voltage is set by first applying all ones (11111111) to the DAC binary input, and then adjusting the 0–2 kΩ potentiometer to the output voltage desired for a full-scale value.

The 1408 (DAC0808) 8-bit D/A converter in Figure 29-1b consists of an R/2R ladder network and produces an output current that is proportional to the magnitude of the binary input. The maximum output current (full-scale output current) is determined by dividing 5 V by the value of resistor R. With the value of R shown in Figure 29-1b, the full-scale output current is 5 mA (5 V/1 kΩ). This is the value of the output current (Io) when all ones (11111111) are applied to the DAC binary input. In order to convert this output current to an output voltage, the 741 op-amp current-to-voltage converter circuit is required. The output voltage (Vo) is equal to the DAC output current (Io) multiplied by the value of resistor R_F. With all ones applied to the DAC binary input, the full-scale output voltage can be adjusted to any desired value to a maximum of 10 V (5 mA × 2 kΩ) by adjusting the value of resistor R_F. The value of the resistor connected to VREF- should be equal to the value of resistor R connected to VREF+ to balance the DAC.

The circuit in Figure 29-2 is an 8-bit voltage output DAC connected to a 7493 wired as a 4-bit binary counter clocked by a 1 kHz pulse generator. Only the low 4-bits of the DAC input are connected to the counter outputs, and the high 4-bits are connected to ground. This means that the DAC will have only 15 steps instead of the full 255 steps of an 8-bit DAC. At the end of the binary count (1111), the counter will reset to zero (0000) and begin a new count sequence. This will cause the DAC output display on the oscilloscope to look like a "staircase" with 15 steps. The resolution (step size) can be determined from the oscilloscope curve plot by measuring the size of one of the steps. The full-scale output voltage (voltage range) can be determined from the oscilloscope curve plot by measuring the maximum DAC output voltage at step 15. The full-scale output voltage will be less than it would have been if the digital input was allowed to go to its full 8-bit count of 255.

The circuit in Figure 29-3 is an 8-bit voltage output DAC connected to two cascaded 74191 4-bit binary counters connected as an 8-bit counter. This will cause the 8-bit DAC to cover the full 0–5 V output range and complete its full 8-bit count of 255 steps. You will notice that this output will look more like a straight line (more like an analog output) than the 15 step curve plot. The larger the number of steps, the more closely the DAC output will come to representing a true analog output (better resolution) for any given voltage range. This circuit can be used to test a DAC to determine if any steps are missing by expanding the oscilloscope horizontal and vertical scales.

Figure 29-1a Digital-to-Analog Conversion

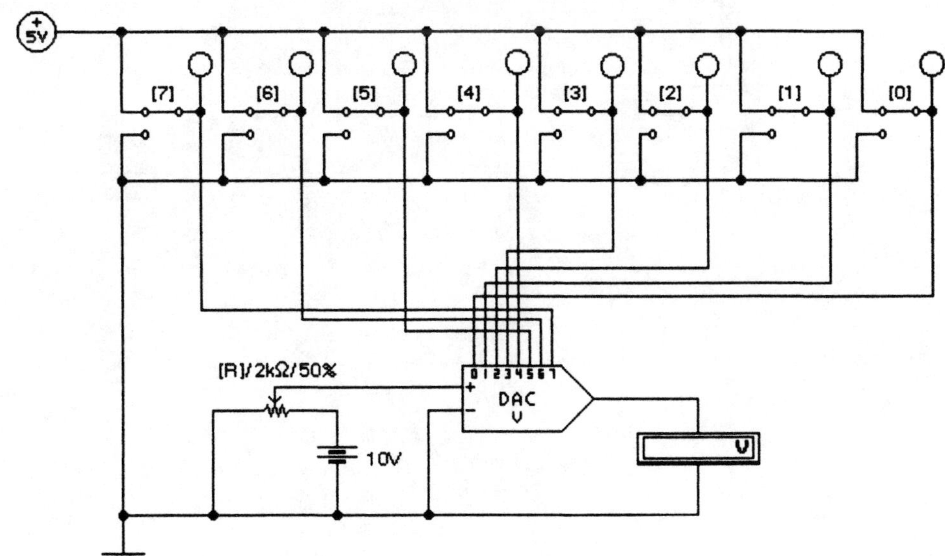

Figure 29-1b Digital-to-Analog Conversion—Hardwired Circuit

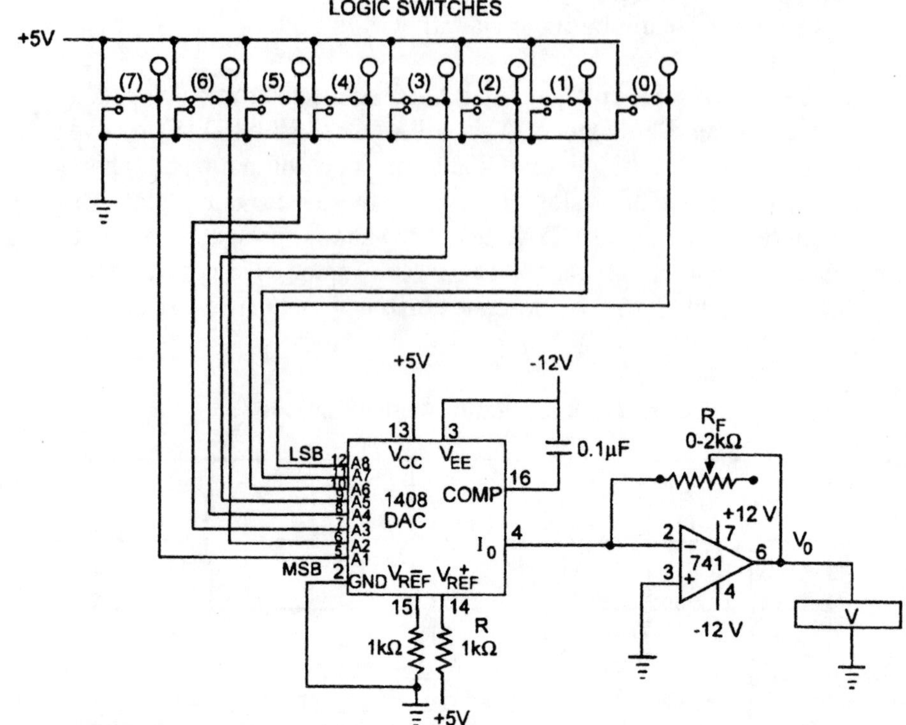

Figure 29-2 DAC Output Waveform (Staircase)

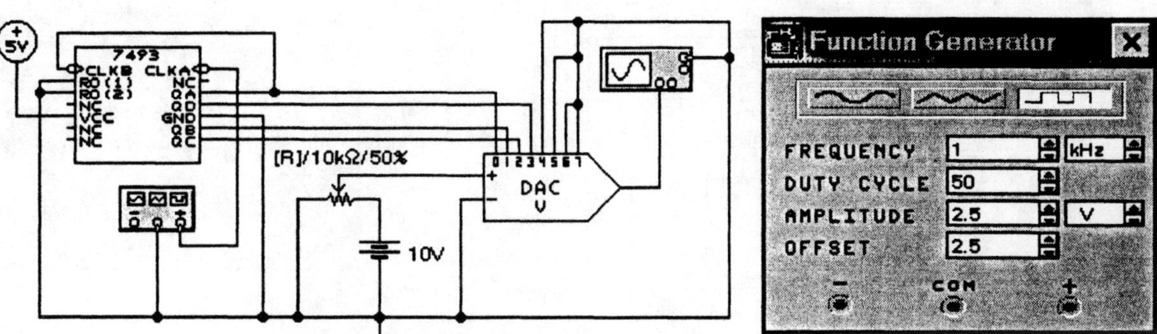

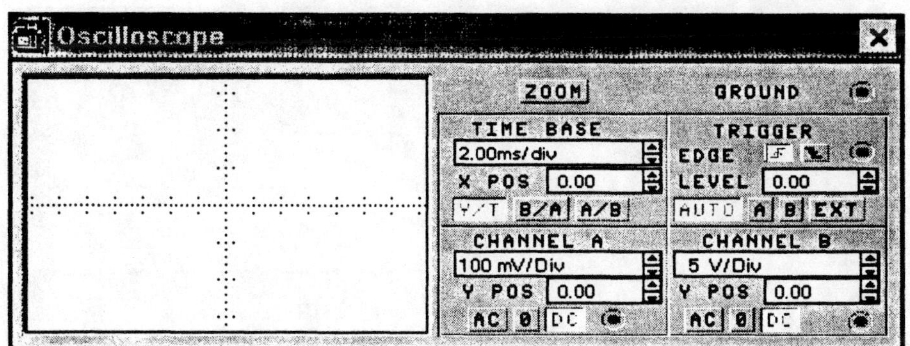

Figure 29-3 DAC Output Resolution

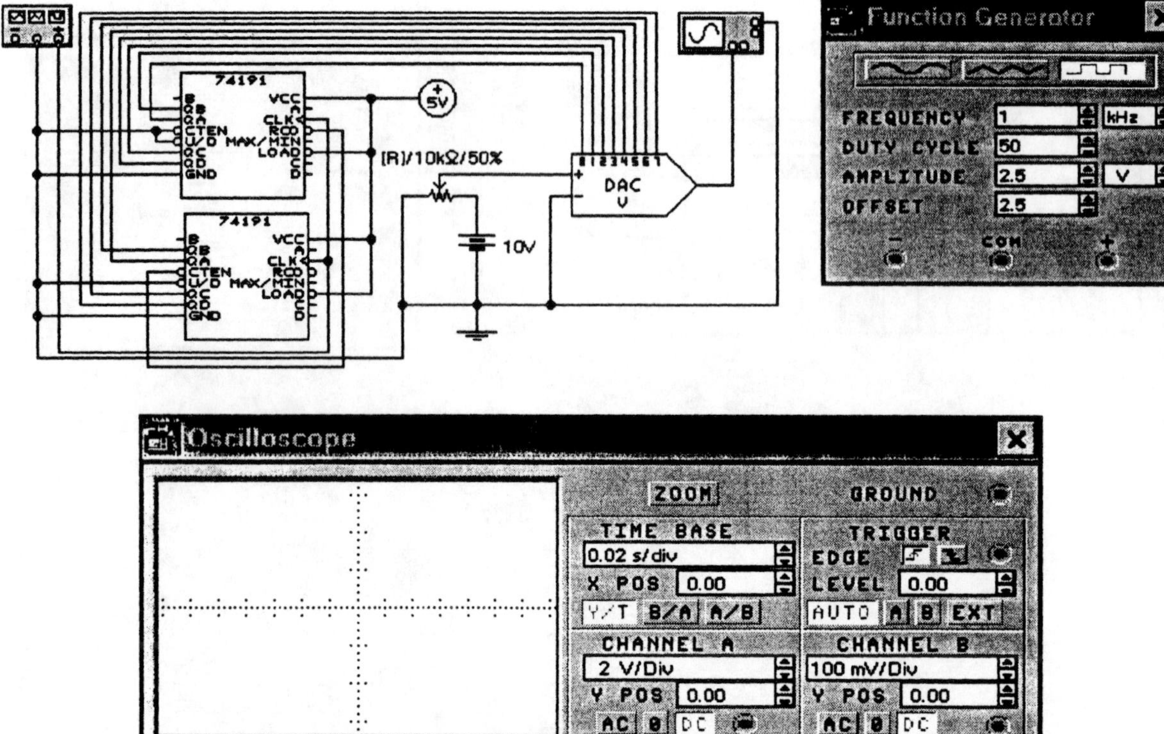

Procedure:

1. Pull down the File menu and open FIG29-1. You are looking at an 8-bit voltage output DAC circuit that will help develop an understanding of the relationship between the digital input and the analog output. The 0–2 kΩ potentiometer is used to set the DAC full-scale output voltage (voltage range). All of the input logic switches should be up (1).

> NOTE: It is recommended that this experiment be performed in a hardwired laboratory. If a hardwired laboratory is available, wire the DAC circuit in Figure 29-1b using a 1408 (DAC0808) current output DAC and a 741 op-amp current-to-voltage converter circuit.

2. With all of the logic switches up (1), click the On-Off switch to run the analysis. Adjust the 0–2 kΩ potentiometer until the DAC output voltage is as close as possible to 5 V. This will set the DAC to a full-scale output voltage of 5 V.

3. By pressing the 0–7 keys on the computer keyboard, change the DAC 8-bit input to binary zero (00000000). Record the DAC output voltage for a binary zero input in Table 29-1.

Table 29-1 DAC Output

Binary Input	Output Voltage
0000 0000	
0000 0001	
0000 0010	
0000 0100	
0000 1000	
1111 1111	

4. Press the 0 key on the computer keyboard to change the DAC 8-bit input to binary one (00000001). Record the DAC output voltage for a binary one input in Table 29-1. Change the logic switches to the remaining binary input values in Table 29-1 and record the DAC output voltages.

Questions: Based on the data in Table 29-1, what is the DAC output offset voltage?

Based on the data in Table 29-1, what is the DAC full-scale output voltage (voltage range)?

Based on the data in Table 29-1, what is the DAC resolution (step size)?

Based on the data in Table 29-1, is the DAC output voltage proportional to the magnitude of the binary input?

5. Based on the DAC full-scale output voltage and the number of steps for an 8-bit input, calculate the expected resolution (step size) of the DAC in Figure 29-1.

Question: How did your calculated resolution (step size) in Step 5 compare with the measured value in Step 4?

6. Based on the data in Table 29-1, calculate the DAC percent resolution.

7. Click the On-Off switch to stop the analysis run. Pull down the File menu and open FIG29-2. You are looking at an 8-bit voltage output DAC connected to a 7493 wired as a 4-bit binary counter clocked by a 1 kHz pulse generator. Only the low 4-bits of the DAC input are connected to the counter outputs, and the high 4-bits are connected to ground. This means that the DAC will have only 15 steps instead of the full 255 steps of an 8-bit DAC. The function generator and oscilloscope settings should be as shown in Figure 29-2. See the Preparation section for more details.

NOTE: If this experiment is being performed in a hardwired laboratory, modify the circuit wired in Step 1 (Figure 29-1b) by removing the logic switches and adding the 7493 counter circuit, function generator, and oscilloscope, as shown in Figure 29-2.

8. Click the On-Off switch to run the analysis. The DAC output shown on the oscilloscope screen is called a "staircase" waveform. Use the oscilloscope "zoom" feature to measure the DAC resolution (step size) and full-scale output voltage (V_{OFS}) and record your answers in the space provided.

Resolution (step size) = _______________

V_{OFS} = _______________

Questions: How did your measured resolution (step size) in Step 8 compare with the value determined in Step 4?

Why was the full-scale output voltage (V_{OFS}) measured in Step 8 different from the value set in Step 2?

9. Click the On-Off switch to stop the analysis run. Pull down the File menu and open FIG29-3. You are looking at an 8-bit voltage output DAC connected to two cascaded 74191 4-bit binary counters connected as an 8-bit counter. This will cause the 8-bit DAC to cover the full 0–5 V output range and complete its full 8-bit count of 255 steps. The function generator and oscilloscope settings should be as shown in Figure 29-3. Click the On-Off switch to run the analysis. Notice how closely the curve plot on the oscilloscope screen resembles a straight line instead of a "staircase" waveform.

NOTE: Unless you have a high speed computer, this curve plot will take a long time to develop.

Questions: Why does the DAC output curve plot on the oscilloscope screen look more like a straight line than a "staircase"?

Based on the oscilloscope curve plot, what is the full-scale output voltage of the DAC?

Name ______________________
Date ______________________

Analog-to-Digital Converters

Objectives:

1 Demonstrate the relationship between the analog input and the digital output of an A/D converter (ADC).
2. Demonstrate how to set the input voltage range of an A/D converter.
3. Investigate ADC conversion and demonstrate how to start conversion.
4. Develop an understanding of the concept of resolution (quantization error) as applied to A/D converters.
5. Demonstrate how to wire a continuous running A/D converter.

Materials:

One 5 V dc voltage supply
One normally open pushbutton switch
Eight logic probe lights
One A/D converter (1-ADC0804 IC)
Two 0–2 kΩ potentiometers
One 15 pF capacitor
One 10 kΩ resistor
Two 0–10 V voltmeters

Preparation:

Most real world physical variables are analog in nature and have values that are within a continuous range. On the other hand, digital systems are discrete in nature and have values that are one of two possibilities. Any analog data that must be input into a digital system must first be converted into digital form. A circuit that converts analog voltages into digital codes is called an analog-to-digital converter (ADC). The magnitude of the digital output is proportional to the analog input voltage. The ADC full-scale input is equal to the analog input voltage that produces the highest binary output (all binary ones). The full-scale input voltage determines the input range of the ADC.

The A/D conversion process is more complex and time consuming than the D/A conversion process because A/D converters require a conversion time before a digital output is obtained. Several important types of A/D converters utilize a D/A converter (DAC) and a comparator as part of their circuitry. The two most common types are the digital ramp ADC (counter-type) and the successive approximation ADC. In the digital ramp ADC, a binary counter output is connected to a DAC input. The DAC output voltage is compared to the analog input voltage by the comparator. When the counter reaches a binary number that causes the DAC output voltage to equal the analog input voltage, the counter stops counting and conversion is complete. The counter binary output is used as the ADC digital output. In the successive approximation ADC, the counter is replaced by a register that is connected to the DAC binary input. Control logic successively modifies the register output until the DAC output voltage is equal to the analog input voltage and conversion is complete. The register binary output is used as the ADC digital output. The successive approximation ADC has a shorter conversion time than the digital ramp ADC and it is one of the most widely used types. The flash ADC has the shortest conversion time, but it requires much more logic circuitry, making it more expensive and complicated to build. Advancements in integrated circuit technology are reducing the cost of the flash ADC.

The resolution (quantization error) of an ADC is defined as the largest analog input voltage change that can occur without producing a change in the digital output. The ADC resolution (quantization error) can be calculated by dividing the full-scale analog input voltage by the number of digital output steps from zero to the maximum count (all binary ones). The number of digital output steps depends solely on the number of ADC output bits, where a 4-bit ADC has 15 output steps and an 8-bit ADC has 255 output steps (0 is not a step). This means that the larger the number of bits, the higher the number of steps and the smaller the resolution (quantization error).

The 8-bit ADC circuit in Figure 30-1a will be used to demonstrate the relationship between the analog input and the digital output for an A/D converter. The ADC full-scale input voltage (input voltage range) is set by adjusting the 0–2 kΩ RANGE potentiometer to the reference voltage (VREF) desired for full-scale input. The analog input voltage (V_{IN}) is set by adjusting the 0–2 kΩ INPUT potentiometer. The input voltage (V_{IN}) can be varied between 0 V and 5 V. ADC conversion is started by pulsing the start-of-conversion input (SOC) to binary one (1) for a short duration. When conversion is complete (the digital output is ready), the end-of-conversion (EOC) output will rise to binary one. The active high output enable (OE) is used to enable the ADC digital output. In Figure 30-1a, it is connected to 5 V (1) to continuously enable the ADC output. The end-of-conversion output (EOC) and the output enable (OE) are used for timing when interfacing the ADC to a microprocessor or computer data bus.

The hardwired ADC0804 in Figure 30-1b is an 8-bit successive approximation ADC that is designed to be interfaced with a microprocessor. It has an internal clock with a frequency determined by the value of R and C. The frequency of the clock controls the length of the ADC conversion time. The ADC0804 has a differential analog input (V_{IN}^+ and V_{IN}^-) which can be used to apply a differential input voltage. The differential input is wired as a single-ended input by connecting V_{IN}^- to ground. The ADC0804 requires a reference voltage that is 1/2 the voltage desired for a full-scale input. If the $V_{REF/2}$ terminal is left open, the full-scale input voltage is equal to V_{cc} (5 V). The ADC0804 also has a digital ground (D_{GND}) and an analog ground (A_{GND}) so that the analog and digital circuits can be isolated from each other. To start conversion, the active low start-of-conversion input (SOC) must be pulsed to binary zero for a short duration. When conversion is complete (the digital output is ready), the active low end-of-conversion (EOC) output will drop to binary zero. The ADC0804 has an active low output enable (RD) and an active low chip select (CS) for interfacing to a microprocessor. In Figure 30-1b, these inputs are connected to ground so that the ADC output is continuously enabled.

Figure 30-1a Analog-to-Digital Conversion

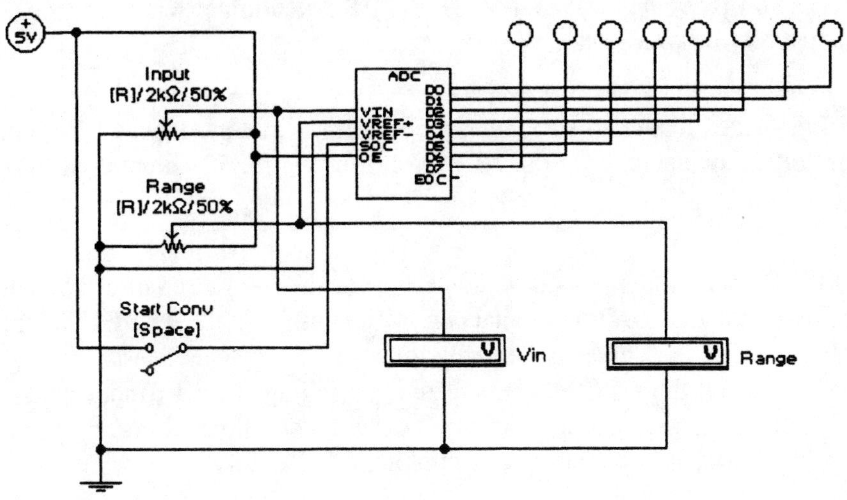

Figure 30-1b Analog-to-Digital Conversion—Hardwired Circuit

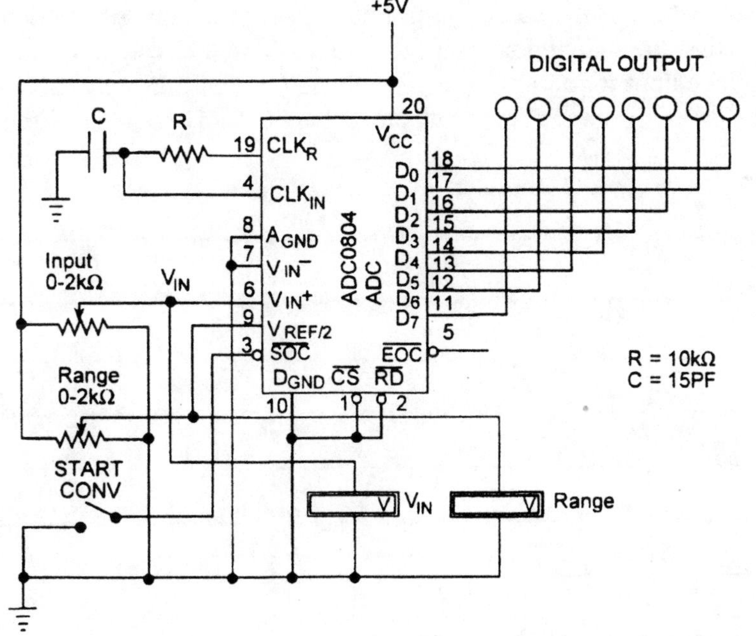

Procedure:

1. Pull down the File menu and open FIG30-1. You are looking at an 8-bit A/D converter circuit that will be used to demonstrate the relationship between the analog input and the digital output. The 0–2 kΩ RANGE potentiometer is used to set the ADC full-scale input voltage (input voltage range). The 0–2 kΩ INPUT potentiometer is used to vary the analog input voltage between 0 V and 5 V.

> NOTE: It is recommended that this experiment be performed in a hardwired laboratory. If a hardwired laboratory is available, wire the ADC circuit in Figure 30-1b using an ADC0804 A/D converter.

2. The START CONV switch should be down (up for the hardwired circuit in Figure 30-1b). Click the On-Off switch to run the analysis. Adjust the 0–2 kΩ RANGE potentiometer until the RANGE voltage is as close as possible to 5 V (2.5 V for the hardwired ADC in Figure 30-1b). This will set the ADC to a full-scale input voltage of 5 V (input voltage range of 0–5 V). The potentiometer setting can be changed by double clicking the potentiometer, changing the % setting in the table, and clicking ACCEPT.

3. Adjust the 0–2 kΩ INPUT potentiometer to a V_{IN} as close as possible to 2.0 V. Press the space bar on the computer keyboard to raise the START CONV switch to binary one (lower to binary zero for the hardwired circuit in Figure 30-1b) to start conversion. Press the space bar again to return the switch down (up for the hardwired circuit in Figure 30-1b). Record the ADC digital output for an analog input voltage (V_{IN}) of 2.0 V in Table 30-1. Repeat the procedure for the remaining values of V_{IN} in Table 30-1 and record the digital outputs.

Table 30-1 ADC Binary Output

V_{IN} (V)	Digital Output	Decimal Equiv
0		
1.0		
2.0		
3.0		
4.0		
5.0		

4. Calculate the decimal equivalent of each binary output in Table 30-1 and record your answers.

Questions: Based on the data in Table 30-1, what is the ADC full-scale input voltage (input voltage range)?

Based on the data in Table 30-1, is the ADC digital output magnitude proportional to the magnitude of the analog input voltage?

5. Based on the data in Table 30-1, calculate the resolution (quantization error) of the ADC in Figure 30-1.

Question: Based on the quantization error, how much can the analog input voltage (V_{IN}) be varied before the digital output changes by one count?

NOTE: Steps 6 and 7 work best in a hardwired laboratory.

6. Click the On-Off switch to stop the analysis run. Connect the EOC output to the SOC input (leave the START CONV switch connected). Click the On-Off switch to run the analysis again. Use the START CONV switch to start conversion. Make sure you return the START CONV switch to its original position. The end-of-conversion (EOC) output will start conversion immediately each time conversion is complete. Now you have a continuous running ADC. Continue varying the analog input voltage (V_{IN}) between 0 V and 5.0 V and observe the digital output. Notice that any changes in V_{IN} will result in an immediate change in the digital output without requiring you to start conversion each time the input voltage is changed. After you have finished varying V_{IN}, go immediately to Step 7.

7. Adjust the 0–2 kΩ RANGE potentiometer until the RANGE voltage is as close as possible to 2.5 V (1.25 V for the hardwired ADC in Figure 30-1b). Adjust the 0–2 kΩ INPUT potentiometer to vary V_{IN} from zero until the digital output just reaches maximum (11111111). Record the analog input voltage for maximum digital output in the space provided. Click the On-Off switch to stop the analysis run. DON'T LEAVE THE ANALYSIS RUNNING.

Analog input voltage = ___________

Question: Based on the result in Step 7, what is the new ADC full-scale input voltage (input voltage range)? Is it what you expected?

Name ______________________
Date ______________________

Data Acquisition

Objectives:

1. Demonstrate how the quality of the analog data being acquired is affected by the relationship between the frequency of the analog data and the sampling rate of the data acquisition system.
2. Demonstrate how the conversion time of the A/D converter in a data acquisition system affects the input data sampling rate.

Materials:

One D/A converter circuit wired in Experiment 29 (Figure 29-1b).
One A/D converter circuit wired in Experiment 30 (Figure 30-1b).
One function generator
One dual-trace oscilloscope

Preparation:

The process of storing analog data as digital data is called data acquisition. In order to store an analog voltage waveshape (voltages that are changing in value with time), an A/D converter must sample the analog voltage frequently so that important voltage values will not be missed. This requires a continuously running ADC that starts conversion immediately after the previous conversion is complete. When the analog data is a fast changing high frequency waveshape, the data sampling points must be close together. This requires an ADC with a short conversion time. If the conversion time is too long (sampling rate too low) for the frequency of the analog waveshape being acquired, the analog reproduction of the digitally stored waveshape will be distorted. The shorter the conversion time (higher the sampling rate), the lower the distortion of the analog reproduction.

The circuit in Figure 31-1 will be used to demonstrate how the relationship between ADC conversion time (sampling rate) and the frequency of the analog input voltage affects the distortion of the analog reproduction of a digital conversion. The digital output of the A/D converter studied in Experiment 30 (Figure 30-1a or Figure 30-1b) is connected to the digital input of the D/A converter studied in Experiment 29 (Figure 29-1a or Figure 29-1b). The ADC end-of-conversion output (EOC) is connected to the ADC start-of-conversion input (SOC) to start conversion immediately after the previous conversion is complete. This will make the ADC run continuously. The conversion time of the ADC will determine the time between sampling points (sampling rate). The analog potentiometer has been replaced by a function generator. The function generator will apply a periodic ramp function to the ADC analog input. The dual trace oscilloscope will monitor the A/D converter analog input and the D/A converter analog output. The relationship between the ADC conversion time (sampling rate) and the frequency of the periodic ramp function will determine how accurately the D/A converter analog output reproduces the A/D converter analog input. If the conversion time (sampling rate) is not fast enough, the D/A converter analog output will not be a good representation of the A/D converter analog input.

Figure 31-1 Data Acquisition

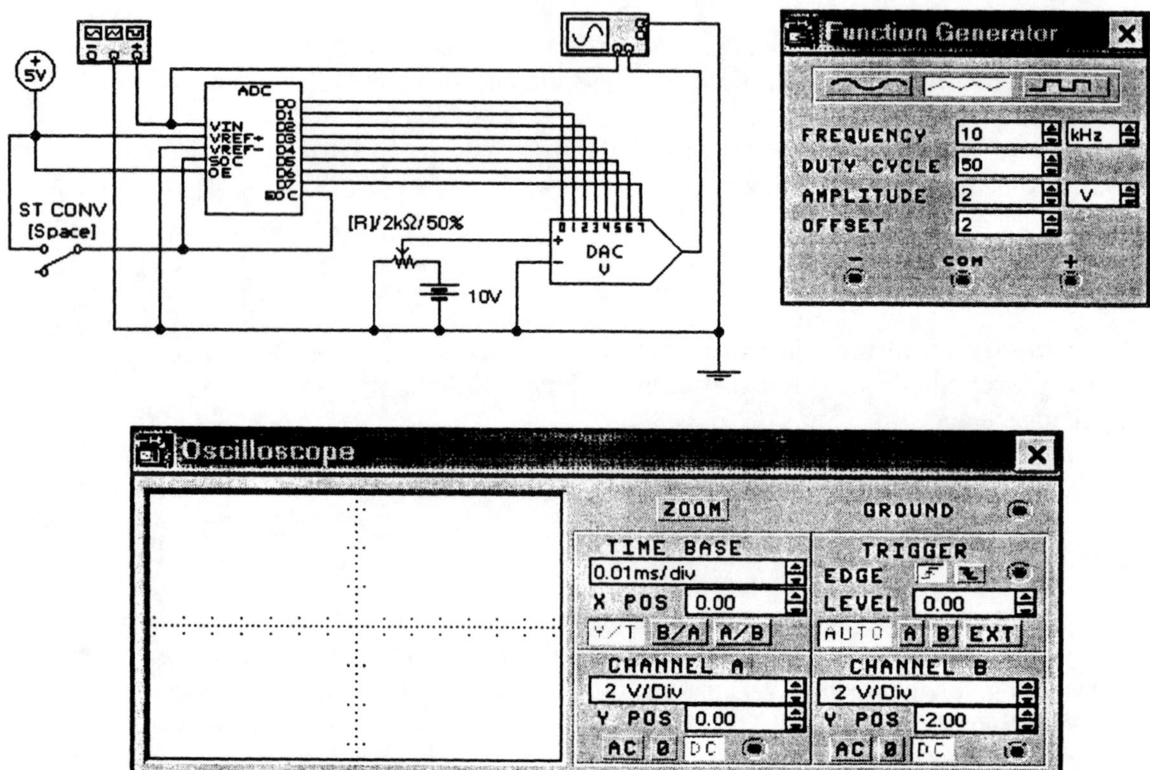

Procedure:

1. Pull down the File menu and open FIG31-1. The digital output of the A/D converter studied in Experiment 30 (Figure 30-1a) is connected to the digital input of the D/A converter studied in Experiment 29 (Figure 29-1a). The ADC end-of-conversion output (EOC) is connected to the ADC start-of-conversion input (SOC) to start conversion immediately after the previous conversion is complete, causing the ADC to run continuously. In the computer simulation, conversion will start automatically when the analysis begins. The voltmeters, the logic probe lights, and the ADC range adjustment potentiometer have been removed and the ADC VREF+ has been connected to 5 V. The analog potentiometer has been replaced by a function generator. The function generator will apply a periodic ramp function to the ADC

analog input. The dual trace oscilloscope will monitor the A/D converter analog input and the D/A converter analog output. The function generator and oscilloscope settings should be as shown in Figure 31-1.

NOTE FOR HARDWIRED LABORATORY USERS: It is recommended that this experiment be performed in a hardwired laboratory. You will use the ADC wired in Experiment 30 (Figure 30-1b) and the DAC wired in Experiment 29 (Figure 29-1b). Make sure that the ADC and the DAC are adjusted for 5 V full-scale. You will connect the digital output of the ADC to the digital input of the DAC. The ADC end-of-conversion output (EOC) is connected to the ADC start-of-conversion input (SOC) to start conversion immediately after the previous conversion is complete, causing the ADC to run continuously. The START CONV switch should remain in the circuit because conversion needs to be initially started in the hardwired circuit. The dc voltmeter that measures V_{IN} should be removed. The logic probe lights should be removed. The input potentiometer should be replaced by a function generator. The function generator will apply a 0–4 V periodic ramp function to the ADC analog input. A dual-trace oscilloscope should monitor the A/D converter analog input and the D/A converter analog output, as shown in Figure 31-1. The function generator and oscilloscope settings should be as shown in Figure 31-1, except the generator frequency should be 1 kHz and the oscilloscope time base should be 0.1 ms/div.

2. Click the On-Off switch to run the analysis. In the computer simulation, conversion will start automatically because the ADC EOC output is connected to the ADC SOC input. (In the hardwired circuit, the START CONV switch must be pushed to start conversion. Make sure you return the switch to its original position after conversion starts). Notice the ADC 10 kHz (1 kHz for the hardwired circuit) periodic input ramp function (red curve plot) and the DAC output (blue curve plot) on the oscilloscope.

Question: Is the DAC output ramp function (blue) a good representation of the ADC input ramp function (red)?

3. Click the On-Off switch to stop the analysis run. Change the frequency of the function generator to 50 kHz (5 kHz for the hardwired circuit). Change the oscilloscope time base to 2.0 μs/div (20 μs /div for the hardwired circuit). Click the On-Off switch to run the analysis again. Notice the ADC 50 kHz (5 kHz for the hardwired circuit) periodic input ramp function (red curve plot) and the DAC output (blue curve plot) on the oscilloscope.

Question: Is the DAC output ramp function (blue) as good a representation of the ADC input ramp function (red) as when the input frequency was 10 kHz (1 kHz for the hardwired circuit)? If not, why not?

4. Use the "zoom" feature on the oscilloscope to measure the time between sampling points and record your answer in the space provided.

   Time = _______________

Question: Based on your answer in Step 4, what is the conversion time of the A/D converter?

Appendix

IC Chip Pin Diagrams

7400

Pin	Left	Right	Pin
1	1A	VCC	14
2	1B	4B	13
3	1Y	4A	12
4	2A	4Y	11
5	2B	3B	10
6	2Y	3A	9
7	GND	3Y	8

7402

Pin	Left	Right	Pin
1	1Y	VCC	14
2	1A	4Y	13
3	1B	4B	12
4	2Y	4A	11
5	2A	3Y	10
6	2B	3B	9
7	GND	3A	8

7404

Pin	Left	Right	Pin
1	1A	VCC	14
2	1Y	6A	13
3	2A	6Y	12
4	2Y	5A	11
5	3A	5Y	10
6	3Y	4A	9
7	GND	4Y	8

7408

Pin	Left	Right	Pin
1	1A	VCC	14
2	1B	4B	13
3	1Y	4A	12
4	2A	4Y	11
5	2B	3B	10
6	2Y	3A	9
7	GND	3Y	8

7410

Pin	Left	Right	Pin
1	1A	VCC	14
2	1B	1C	13
3	2A	1Y	12
4	2B	3C	11
5	2C	3B	10
6	2Y	3A	9
7	GND	3Y	8

7411

Pin	Left	Right	Pin
1	1A	VCC	14
2	1B	1C	13
3	2A	1Y	12
4	2B	3C	11
5	2C	3B	10
6	2Y	3A	9
7	GND	3Y	8

7420

Pin	Left	Right	Pin
1	1A	VCC	14
2	1B	2D	13
3	NC	2C	12
4	1C	NC	11
5	1D	2B	10
6	1Y	2A	9
7	GND	2Y	8

7421

Pin	Left	Right	Pin
1	1A	VCC	14
2	1B	2D	13
3	NC	2C	12
4	1C	NC	11
5	1D	2B	10
6	1Y	2A	9
7	GND	2Y	8

7427

Pin	Left	Right	Pin
1	1A	VCC	14
2	1B	1C	13
3	2A	1Y	12
4	2B	3C	11
5	2C	3B	10
6	2Y	3A	9
7	GND	3Y	8

7432

Pin	Left	Right	Pin
1	1A	VCC	14
2	1B	4B	13
3	1Y	4A	12
4	2A	4Y	11
5	2B	3B	10
6	2Y	3A	9
7	GND	3Y	8

7442

Pin	Left	Right	Pin
1	0	VCC	16
2	1	A	15
3	2	B	14
4	3	C	13
5	4	D	12
6	5	9	11
7	6	8	10
8	GND	7	9

7448

Pin	Left	Right	Pin
1	B	VCC	16
2	C	0F	15
3	LT′	0G	14
4	BI/RB0′	0A	13
5	RBI′	0B	12
6	D	0C	11
7	A	0D	10
8	GND	0E	9

7474

Pin	Left	Right	Pin
1	1CLR′	VCC	14
2	1D	2CLR′	13
3	1CLK	2D	12
4	1PRE′	2CLK	11
5	1Q	2PRE′	10
6	1Q′	2Q	9
7	GND	2Q′	8

7475

Pin	Left	Right	Pin
1	1Q′	1Q	16
2	1D	2Q	15
3	2D	2Q′	14
4	3C,4C	1C,2C	13
5	VCC	GND	12
6	3D	3Q′	11
7	4D	3Q	10
8	4Q′	4Q	9

7483

Pin	Left	Right	Pin
1	A4	B4	16
2	Σ3	Σ4	15
3	A3	C4	14
4	B3	C0	13
5	VCC	GND	12
6	Σ2	B1	11
7	B2	A1	10
8	A2	Σ1	9

7486

Pin	Left	Right	Pin
1	1A	VCC	14
2	1B	4B	13
3	1Y	4A	12
4	2A	4Y	11
5	2B	3B	10
6	2Y	3A	9
7	GND	3Y	8

7490

Pin	Left	Right	Pin
1	CKB	CKA	14
2	R01	NC	13
3	R02	QA	12
4	NC	QD	11
5	VCC	GND	10
6	R91	QB	9
7	R92	QC	8

7493

Pin	Left	Right	Pin
1	CKB	CKA	14
2	R01	NC	13
3	R02	QA	12
4	NC	QD	11
5	VCC	GND	10
6	NC	QB	9
7	NC	QC	8

74112

Pin	Left	Right	Pin
1	1CLK	VCC	16
2	1K	1CLR′	15
3	1J	2CLR′	14
4	1PRE′	2CLK	13
5	1Q	2K	12
6	1Q′	2J	11
7	2Q′	2PRE′	10
8	GND	2Q	9

74121

Pin	Left	Right	Pin
1	$\overline{Q}$	VCC	14
2	NC	NC	13
3	A1	NC	12
4	A2	Rext/Cext	11
5	B	Cext	10
6	Q	Rint	9
7	GND	NC	8

74138

Pin	Left	Right	Pin
1	A	VCC	16
2	B	Y0	15
3	C	Y1	14
4	G2A′	Y2	13
5	G2B′	Y3	12
6	G1	Y4	11
7	Y7	Y5	10
8	GND	Y6	9

74147

Pin	Left	Right	Pin
1	4	VCC	16
2	5	NC	15
3	6	D	14
4	7	3	13
5	8	2	12
6	C	1	11
7	B	9	10
8	GND	A	9

74148

Pin	Left	Right	Pin
1	4	VCC	16
2	5	E0	15
3	6	GS	14
4	7	3	13
5	E1	2	12
6	A2	1	11
7	A1	0	10
8	GND	A0	9

74151

Pin	Left	Right	Pin
1	D3	VCC	16
2	D2	D4	15
3	D1	D5	14
4	D0	D6	13
5	Y	D7	12
6	W	A	11
7	G′	B	10
8	GND	C	9

74157

Pin	Left	Right	Pin
1	A/B′	VCC	16
2	1A	G′	15
3	1B	4A	14
4	1Y	4B	13
5	2A	4Y	12
6	2B	3A	11
7	2Y	3B	10
8	GND	3Y	9

74173

Pin	Left	Right	Pin
1	M	VCC	16
2	N	CLR	15
3	1Q	1D	14
4	2Q	2D	13
5	3Q	3D	12
6	4Q	4D	11
7	CLK	G2′	10
8	GND	G1′	9

74190

Pin	Left	Right	Pin
1	B	VCC	16
2	QB	A	15
3	QA	CLK	14
4	CTEN′	RCO′	13
5	D/U′	MAX/MIN	12
6	QC	LOAD′	11
7	QD	C	10
8	GND	D	9

74191

Pin	Name	Name	Pin
1	B	VCC	16
2	QB	A	15
3	QA	CLK	14
4	CTEN'	RCO'	13
5	D/U'	MAX/MIN	12
6	QC	LOAD'	11
7	QD	C	10
8	GND	D	9

74194

Pin	Name	Name	Pin
1	CLR'	VCC	16
2	SR	QA	15
3	A	QB	14
4	B	QC	13
5	C	QD	12
6	D	CLK	11
7	SL	S1	10
8	GND	S0	9

74280

Pin	Name	Name	Pin
1	G	VCC	14
2	H	F	13
3	NC	E	12
4	I	D	11
5	EVEN	C	10
6	ODD	B	9
7	GND	A	8

555

Pin	Name	Name	Pin
1	GND	VCC	8
2	TRI	DIS	7
3	OUT	THR	6
4	RES	CON	5

Appendix

B Notes on Using Electronics Workbench

1. If you wish to remove a component from a circuit, disconnect both terminals from the circuit; otherwise you may get an error message.

2. You can change a component value by double-clicking it with the mouse and changing the menu value using the keyboard.

3. You can bring down an instrument enlargement by double-clicking the instrument with the mouse.

4. The circuit disk provided with this manual is write protected; therefore, you cannot save a changed circuit to the disk. If you wish to save a changed circuit, you must select "Save As" in the File menu and save it on another disk or the hard drive.

5. The color of a logic analyzer or oscilloscope curve trace is the same as the color of the circuit wire connected to the input. A wire color can be changed by double-clicking the wire with the mouse and selecting a new color from the table on the screen.

6. In EWB version 4.1 or higher, wires can be moved by placing the arrow on the wire, clicking the left mouse button, and dragging the wire to a new position.

Bibliography

Floyd, T. L. *Digital Fundamentals*. 6th ed. Upper Saddle River, NJ: Prentice Hall, 1997.

Klietz, W. *Digital Electronics: A Practical Approach*. 4th ed. Upper Saddle River, NJ: Prentice Hall, 1996.

Leach, D. P., and Malvino, A. P. *Digital Principles and Applications*. 5th ed. New York: McGraw-Hill, 1995

Tocci, R. J., and Widmer, N. S. *Digital Systems: Principles and Applications*. 7th ed. Upper Saddle River, NJ: Prentice Hall, 1998.